PHYSICAL GEOLOGY
LAB EXPLORATION

SECOND EDITION

X. MARA CHEN | THOMAS CAWTHERN

Kendall Hunt
publishing company

Front cover image: Kilauea lava meets the sea at the Kamokuna cliff, in Hawaii Volcanoes
National Park provided by Mara Chen

Back cover image: Kilauea caldera in Hawaii Volcanoes National Park provided by Mara Chen

Interior images provided by Mara Chen (unless otherwise noted).

www.kendallhunt.com
Send all inquiries to:
4050 Westmark Drive
Dubuque, IA 52004-1840

Copyright © 2015, 2018 by Kendall Hunt Publishing Company

ISBN 978-1-5249-5187-0

Published in the United States of America

CONTENTS

PREFACE

This book is designed and written for the introductory physical geology course. It provides students with fundamental concepts and principles, as well as examples and lab exercises to apply what is learned in lectures and reading. The primary objectives are to promote and nurture learning through critical thinking and active hands-on activities for problem solving, to achieve a better understanding of earth's internal composition and structures, and to obtain a deeper appreciation for the surface landform features and earth's external dynamic processes.

Any learning process starts with curiosity and questioning; proceeds with exploration, critical thinking, analysis, and comprehension; and then arrives at understanding and the appreciation of the beauty of applying that knowledge. The reward of discovering something new about the world and ourselves leads us to more meaningful and fulfilling learning experiences.

Earth, the home for life, serves as an excellent laboratory. As soon as we step outside, nature becomes our natural classroom without a roof, walls, and borders. Beneath the earth's vast surface, grand, internal dynamic processes help shape and reshape its surface landforms. As the saying goes, "If you dig a hole deep enough, you will reach China." We have not been able to dig or drill through the entire earth, nor will we any time soon. Yet, this does not take our pleasure away from imagining the possible pathways to its core and exploring the earth's inner making indirectly.

Our imagination is clearly far more penetrating than our physical abilities and the capabilities of the tools we use. Our scientific imagination and curiosity lead us to many questions such as: What is the earth made of? What inner structures do we expect to see? Why does the earth look the way it does? Why are certain locations more prone than others to natural disasters such as earthquakes and volcanoes? How did majestic mountain ranges come to be? Will mountains remain the same height and oceans the same depth over time?

To address these and other questions, this manual is organized into ten major labs. Each lab is organized around a central topic. The lectures of the course are generally divided into three major components: (1) earth's composition, (2) geological time and internal dynamic processes, and (3) external processes and surface landforms. The lab exercises are grouped into two major parts: earth composition and dynamic processes. Together, this course provides a general overview of what the earth is made of and how the earth system operates.

From a chemist's point of view, the world is made of atoms. In a physicist's eyes, the universe consists of different systems that interact through different forms of energy. From a geologist's vantage point, the earth is made of rocks and minerals, and the earth system is operated by both internal and external engines.

Needless to say, earth is a dynamic system with its interior neither homogeneous nor constant or consistent across space and over time. This introductory course will present a general picture and give us some level of overall understanding of the earth's composition, which provides a foundation for examining its dynamic processes and associated landform features.

Minerals and rocks are the fundamental alphabet blocks of the earth. Their basic physical and chemical properties help us better understand different geologic narratives of the earth's evolution spanning 4.6 billion years of geologic time. There are more than 100 known elements, yet only eight elements make up 99% of the earth's crust. The study of the chemical makeup of rocks and minerals as well as an understanding of the physical processes shaping the surface of the earth facilitates our understanding of this dynamic planet we call home. For example, quartz is one of the most durable minerals in the world, and the beach sands often have had some of the longest journeys across the land before reaching the sea. Diamond, on other hand, might have completed some of the roughest vertical journeys from great depths to where they are found and mined.

Beyond the most precious resources to man and other forms of life, the value and beauty of the earth are created and defined by the earth's dynamic processes. These processes can be both internal and external. In general, the internal processes tend to erect the landscape, while the external processes try to denudate the differences. Several labs focus on earth's geological time, major geological structures, fluvial processes, coastal processes, and glacial systems.

To make the lab learning experiences more effective, students are expected to read and review the lab and related course notes before coming to the lab, work as a group during the lab, and review and enhance understanding of each lab in wrap-up sessions.

Acknowledgments

We want to extend our thanks to our students for showing their interests in geology, to our colleagues Dr. Brent Skeeter and Dr. Brent Zaprowski for offering their valuable professional suggestions, and to our families for their support and love.

LAB ONE

MINERALS

LEARNING OBJECTIVES:

- Identify minerals in products we use every day, and begin to associate these applications with the different properties of minerals.
- Learn and apply diagnostic tests to examine a mineral's properties.
- Learn to distinguish minerals from nonmineral materials.
- Learn to distinguish common minerals from one another using their various properties.
- Familiarize yourself with the possible links between earth's materials and earth's dynamic processes.

Why Should We Study Minerals?

Minerals are the fundamental building blocks of the earth's landscape. They are essential to our daily life and have a wide range of applications, including cosmetics, jewelry, construction products, fertilizers, food, and medicine, to name a few. Studying minerals is the first step to understanding rocks, landscapes, and the dynamic processes operating on the interior and exterior of the earth. Importantly, these factors have had a significant role in the establishment, development, thrive, and failure of human civilizations.

Definitive Properties of Minerals

Most major rock-forming minerals may be distinguished from one another by using a common set of properties. These common properties, often referred to as definitive properties, are also used to identify minerals from other common materials. The following criteria are used in distinguishing a mineral from other forms of matter.

Minerals MUST be/have:

1. Naturally occurring or naturally formed solids

2. Inorganic chemical compounds

3. A definite chemical composition (this may be strictly fixed or it may vary within a well-defined limit), regardless of where they are formed or found

4. An orderly internal crystalline structure.

Apply What You Learned

Determine which of the following materials fit the definition of a mineral, and explain why, if it is not a mineral.

1. Is diamond a mineral? ❏ **Yes** ❏ **No** _____

2. How about synthetic diamond? ❏ **Yes** ❏ **No** _____

3. Snowflakes? ❏ **Yes** ❏ **No** _____

4. Window glass? ❏ **Yes** ❏ **No** _____

5. Fossils? ❏ **Yes** ❏ **No** _____

6. Coal? ❏ **Yes** ❏ **No** _____

Mineral Formation

Each mineral has a unique internal crystalline texture, which, in a way, reflects a unique history of formation. The growth of mineral crystals occurs as atoms are combined and bonded together to form a well-defined, three-dimensional lattice.

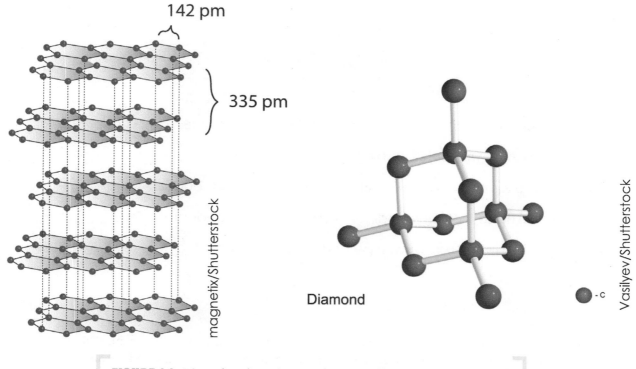

FIGURE 1.1 Atomic structure of graphite and diamond

Minerals commonly form in two general ways: either by precipitating out of solution or by crystallizing from a magma or lava as it cools. The chemical composition and the physical environment (temperature, pressure, and space) affect how a mineral grows. Together, they dictate the size and chemical bonding structure of a mineral.

Apply What You Learned

Use your lecture notes or textbook to answer the following questions:

1. Give an example of a mineral formed under high temperature and pressure from melt.

   _____

2. Give an example of a mineral formed from precipitation out of a solution.

   _____

Chemical Bonding

Minerals, like all matter, are made up of atoms, which are chemically bonded to one another to form crystals. There are four common chemical bonds in minerals—ionic, covalent, metallic, and the Van Der Waals. Many mineral properties depend on a mineral's atomic bonding. Moreover, the bond type is dictated by the type and number of atoms present in a particular mineral.

Apply What You Learned

Use your lecture notes or textbook to answer the following questions:

1. Name a mineral with ionic bonding. _____

2. Provide an example of a mineral that contains covalent bonds. _____

3. Minerals that have metallic bonding tend to be good conductors. True_____ False_____

4. The mineral graphite has what type of chemical bond? _____

Chemical Composition

Most minerals are chemical compounds, consisting of two or more elements. Other minerals, such as diamond, may consist entirely of one element (carbon). Of the more than 100 known elements, 99% of earth's crust is made up of only eight elements (O, Si, Al, Fe, Ca, Mg, Na, K). The two most abundant elements, silicon (Si) and oxygen (O), comprise over 83% by volume of the composition of the crust.

TABLE 1.1 Relative abundance of the top eight elements in earth's crust

Element	Symbol	Weight%	Volume %
Oxygen	O	46.60	62.55
Silicon	Si	27.72	21.22
Aluminum	Al	8.13	6.47
Iron	Fe	5.00	1.92
Calcium	Ca	3.63	1.94
Sodium	Na	2.83	2.64
Potassium	K	2.59	1.42
Magnesium	Mg	2.09	1.84
		98.59	100

Common Mineral Groups

The Periodic Table has been constructed to group elements with similar properties. Minerals are grouped together according to their chemical makeup. Grouping minerals in this manner allows us to categorize minerals based on specific properties of each mineral within the group. The following is a list of the major rock-forming mineral groups:

1. Silicates ($\sim SiO_4)^{4+}$: make up more than 95% of earth's crust. Examples include quartz (SiO_2), diopside ($CaMgSi_2O_6$)—a pyroxene, albite ($NaAlSi_3O_8$)—a feldspar.

2. Carbonates ($\sim CO_3$): effervesces in diluted hydrochloric acid. Common minerals include calcite ($CaCO_3$) and dolomite ($CaMgCO_3$).

3. Oxides ($\sim O$): magnetite (Fe_3O_4), corundum (Al_2O_3), hematite (Fe_2O_3).

4. Sulfides ($\sim S$): pyrite (FeS_2), Chalcopyrite($Cu,Fe)S_2$, Galena (PbS).

5. Sulfates ($\sim SO_4$): anhydrite ($CaSO_4$), barite ($BaSO_4$), gypsum ($CaSO_4 * 2H_2O$).

6. Phosphates ($\sim PO_4$): apatite [$Ca_5(PO_4)_3OH$].

7. Halides: principle anions are halogens. Examples include: halite ($NaCl$)—commonly called table salt, fluorite (CaF_2).

8. Native elements: entirely made of one element, such as gold (Au), silver (Ag), copper (Cu), diamond (C), and graphite (C).

Apply What You Learned

---

1. Silicate minerals comprise the most abundant mineral group on earth. What does this statement imply about how easily these minerals can be physically or chemically degraded?

2. Do all silicate minerals react the same way to the physical or chemical changes? Why? Or why not?

Diagnostic Properties of Minerals

There are estimated to be more than 2,000 known minerals. Mineral identification is both fascinating and challenging. The common diagnostic properties for mineral identification include physical appearance (shape, color, streak, luster, density), minerals' responses to applied forces (fracture, cleavage, hardness, magnetism), chemical properties (taste, chemical reactions), and other unique properties (fluorescence, degree of transparency to light, double refraction, magnetism, etc.).

Crystal Form or Crystal Habit

Crystal form or crystal habit refers to the external geometric shape of a mineral crystal. Perfect crystals are rare due to uniquely restrictive environments in which these crystals form naturally. Still, many mineral crystals develop into beautiful geometrical shapes, given unrestricted environmental conditions. Some minerals can even grow into different geometric shapes (called polymorphs), leading to the formation of new minerals with the same chemical compositions (e.g., diamond versus graphite).

Common crystal forms:

1. Cubic: cube shaped (e.g., halite, galena)

2. Octahedral: octahedrons (e.g., fluorite)

3. Tabular: rectangular or trapezoidal (e.g., feldspar)

4. Granular: approximately equal length (e.g., olivine, garnet)

5. Acicular: slender or needlelike crystals (e.g., rutile)

6. Prismatic: prismatic faces (e.g., quartz)

7. Reniform: kidney shaped (e.g., malachite, hematite)

FIGURE 1.2 Different crystal forms (Credit: Mara Chen and Tom Cawthern)

Color

The color of a mineral is probably one of the most widely used properties in mineral identification, even though color alone is not very reliable. A mineral's color depends on how it reflects visible light. Just as very slight variations in the ingredients in your favorite carbonated beverage may result in drastically different flavors between brands, a mineral can appear in a variety of colors, owing in large part to slight variations in its chemical composition (crystalline structure), inclusions, impurities, and radiation background.

The most common ions that often cause color variations in minerals are the transition metals Ti, V, Cr, Mn, Fe, Co, Ni, and Cu. For example, the mineral quartz can be colorless, clear, milky white, smoky gray, rosy pink, and/or citrine (orange) due to slight additions of these chemicals in the mineral structure.

Streak

Streak is the color of a mineral in its powdered or pulverized form. Although the color of a mineral may vary, the color of a mineral's streak is often fairly constant, so the streak of a mineral can differ from its color. Streak is obtained by rubbing a mineral against an unglazed porcelain plate with a hardness (H ~6.5).

Apply What You Learned

---

1. Do all minerals have a streak using the unglazed porcelain plate? Why? Or why not?

2. Are a mineral's streak and its color always the same?

Luster

Luster is the general appearance of a mineral's refectivity of light. It can shine like a metal surface or it may reflect light in a nonmetal way, much like a pane of glass will reflect light.

- **Metallic luster**: used to describe those minerals that are opaque (will not transmit light through—are not transparent). Generally speaking, these minerals have a high abundance of metallic elements and therefore appear as a metal.

- **Nonmetallic luster**: describes the varied appearance of minerals that do not outwardly resemble metal. This luster may range from brilliant and shining to dull and earthy (poor reflection). Common nonmetallic lusters include adamantine, vitreous/glassy, resinous, greasy, pearly, silky, or earthy.

Apply What You Learned

Use the images in Figure 1.3 to answer the following questions:

1. Is the mineral shown in the left-hand photo metallic or nonmetallic?

2. Is there any metallic mineral in the center photo?

3. What types of luster can you see in the right-hand photo?

Cleavage and Fracture

Cleavage

Beyond the most common external physical properties of a mineral (shape and color), cleavage is one of the most important internal properties. Cleavage is a mineral's tendency to break along planes of weakness in its crystal lattice, forming smooth and orderly surfaces. These surfaces reflect the weak microscopic bonds holding elements together in the crystal lattice, and are therefore helpful in determining a mineral's identity.

Cleavage planes may occur in several different directions, or orientations. Some cleavage planes may be more developed than others that are oriented in a different direction. Because of this, cleavage planes are often classified according to how well developed or distinct these flat surfaces appear: perfect, good, and poor.

Cleavage planes are often confused with the flat surfaces that are the result of a mineral's geometrical crystal shape. Ideally, the easiest way to differentiate the two is to break the mineral in question with a rock hammer. If the mineral breaks into a repeating geometrical shape different than the original mineral, then the first shape was the crystal shape and the second shape is the result of intersecting cleavage planes. We do not have the luxury of breaking minerals with a rock hammer. Instead, the easiest way to distinguish cleavage planes from crystal habit is to look at a number of different samples of the same mineral. If you are able to recognize a repeatable pattern in the shape of a mineral as the size varies from large to very small, then the shape is likely the result of intersecting cleavage planes.

Mineral Cleavage	Direction	Angle	Example
	One direction	n/a	Biotite 
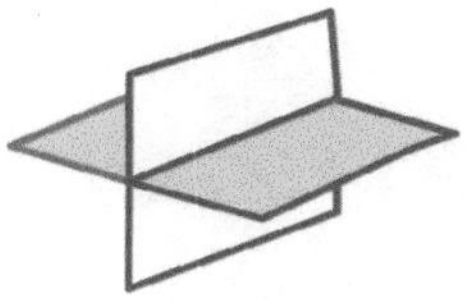	Two directions	=90°	K-feldspar 
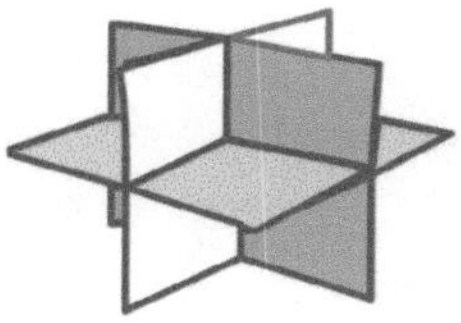	Three directions	=90°	Galena 
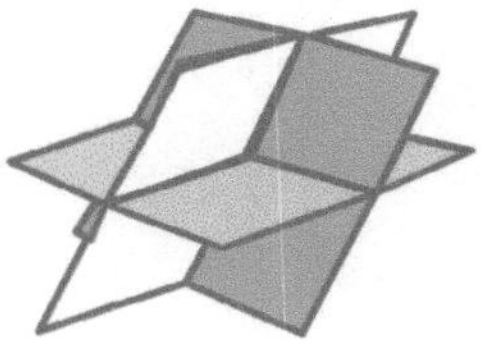	Three directions	≠90°	Calcite 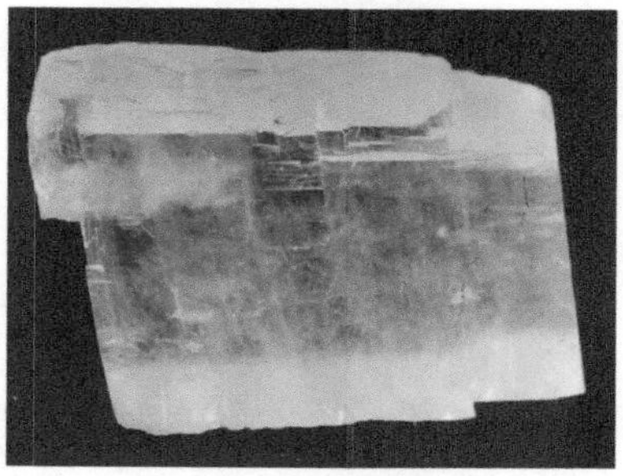

FIGURE 1.4 The directions and angles of mineral cleavage (Credit: Mara Chen)

Fracture

Fracture is another response a mineral may exhibit when it is broken. In some mineral crystals, the strength of the bonds are approximately the same in all directions. As a result, these crystals often break along irregular surfaces. There are different kinds of fractures: conchoidal, fibrous, irregular. Sometimes, fracture surfaces may appear to be flat and even surfaces, resembling cleavage surfaces, but are not repeatable throughout the sample.

It is important to note that all minerals can break and form fractures, but only some minerals have cleavages.

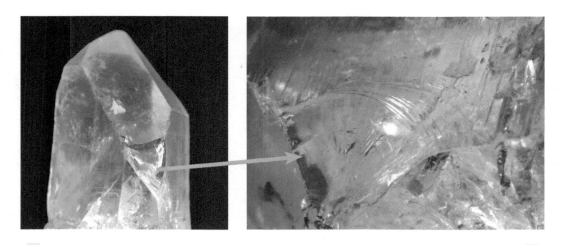

FIGURE 1.5 Conchoidal fracture (Credit: Mara Chen & Tom Cawthern)

Hardness

Many minerals are highly valued for their hardness in industrial applications, as well as their metaphorical symbol of everlasting love. Hardness is a mineral's resistance to scratching or abrasion. A standard relative scale, developed by Friedrich Mohs in 1812, contains standard minerals with a hardness scale ranging from 1 to 10.

TABLE 1.2 Moh's Hardness Scale of Minerals

Mineral Name	Moh's Hardness Scale	Common Test Objects
Talc	1	
Gypsum	2	Fingernail (2.5)
Calcite	3	Copper Penny (3 -3.5)
Fluorite	4	
Apatite	5	Glass Plate (5.5)
Feldspar	6	
Quartz	7	Streak Plate (6.5-7)
Topaz	8	
Corundum	9	
Diamond	10	

Hardness is not a simple linear scale—for example, diamond is not 10 times harder than talc (see the absolute Knoops scale chart in Figure 1.6).

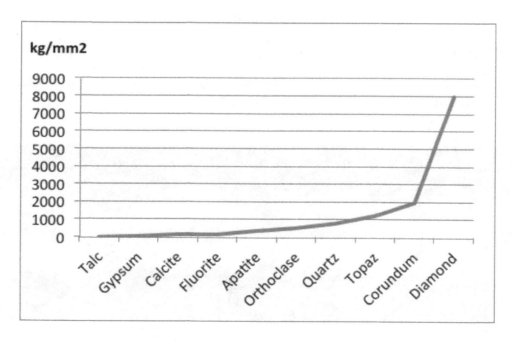

FIGURE 1.6 The absolute Knoops scale of mineral hardness
(data source: http://www.minsocam.org/msa/collectors_corner/arc/knoop.htm)

A hardness test is performed by scratching standard minerals or materials of known hardness with an unknown mineral sample. For example, your fingernails (H ~2.5), a copper penny (H ~3), a glass plate (H ~5.5), and a steel file (H ~6) are easily accessible in most circumstances, and can easily be used to determine the hardness of a mineral.

Some minerals may exhibit a range of hardnesses depending on the direction along which you scratch it. For example, calcite has a hardness of 3 for all surfaces except one, which has a slightly softer hardness of 2.

Density and Specific Gravity

The density of a mineral is defined as the mass of that mineral (typically measured in grams) per unit volume (typically measured in cubic centimeters, cm^3). Specific gravity is the dimensionless ratio between the density of a mineral to the density of water. Most rock-forming silicate minerals have specific gravities ranging from 2.5 to 3.5. Specific gravity is often favored over measurements of density because some minerals have unique values of specific gravity. For example, galena 7.5, cinnabar 8.1, silver 10.5, gold 19.3, and platinum 21.5.

Tenacity

Tenacity is a mineral's resistance to breaking, crushing, or bending. Mineral tenacity can range from brittle, malleable, ductile, flexible, to elastic.

Special Properties

Some minerals have special properties, such as magnetism (magnetite), a unique taste (halite), double refraction (calcite), reaction to HCl acid (calcite), and fluorescence.

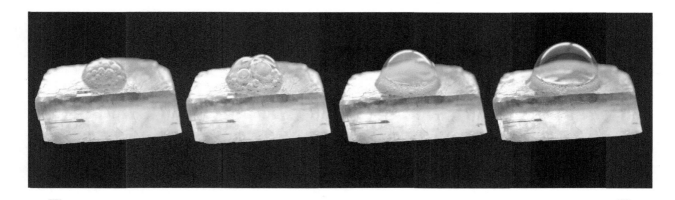

FIGURE 1.7 Calcite reaction with HCl acid (Credit: Mara Chen & Tom Cawthern)

Mineral Identification Process

Mineral identification often takes into consideration a combination of the properties discussed above. Some minerals are easier to identify than others; you will find, however, that all of the major rock-forming minerals will become easier to identify with practice. Mineral identification is a process that takes an initial time investment that pays dividends when we study the three main groups of rocks: igneous, metamorphic, and sedimentary. The flowchart in Figure 1.8 is commonly used as a guide in mineral identification in introductory geology courses. After following the flowchart, you can further test other properties (see Tables 1.3–1.5) to identify specific mineral samples.

In this lab, you are provided with a tray of 12 mineral samples. Please follow the mineral flowchart to record your observations in the worksheet and correctly identify each of the minerals provided.

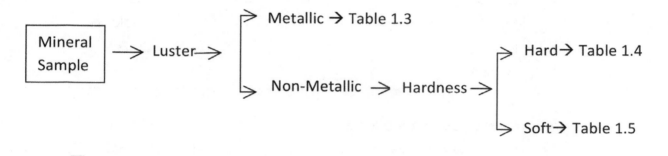

FIGURE 1.8 Mineral identification flowchart (Credit: Mara Chen)

TABLE 1.3 Metallic minerals

Streak	Other Diagnostic Properties	Name Composition
Dark colors	Steel gray, dark gray, black, slippery, greasy Massive mass H = 1-2 D = 2.1-2.3	Graphite C
	Copper yellow, may have purplish hue Massive mass H = 4 D = 4.3	Chalcopyrite Cu,FeS_2
	Silver gray Perfect Cubic cleavage Heavy: D = 7.6, silver gray H = 2.5	Galena PbS
	Black, dark gray D = 5.2 Hard: H = 6	Magnetite Fe_3O_4
	Brass yellow Cubic crystal form or random massive granular aggregates Striations on crystal surface	Pyrite FeS_2
Brown, reddish brown	Steel gray or reddish brown Massive or plate shape Uneven surface fractures	Hematite Fe_2O_3
Yellow, light brown	Yellow, light brown, may have black streak Massive mass Hard: H = 5 – 5.5 D = 3.5 – 4	Limonite $Fe_2O_3*H_2O$

TABLE 1.4 Nonmetallic minerals (harder than glass)

Cleavage	Other Diagnostic Properties	Name Composition
Prominent	Dark green, brown, or black H = 6; D = 3.4 Long bar-shaped, prismatic 6-sided crystals Two-direction cleavage at 60 and 1200	Amphibole Group Na, Ca, Mg, Fe, Al silicates
	Light to dark pink, peach, or salmon color H = 6-6.5; D = 2.5 2-direction perfect to good cleavage	Potassium Feldspar $KAlSi_3O_8$
	White, light gray to dark bluish gray H = 6-6.5 Good cleavage in two directions	Plagioclase Feldspar $NaAlSi_3O_8$ $CaAlSi_3O_8$
	Dark green to black Prismatic crystals H = 6, D = 3.5 Good cleavage in two directions	Pyroxene Group Ca, Mg, Fe, Al Silicates
Absent	Red, dark reddish brown, yellow, or even blue H = 7-7.5; D = 3.5-4.5 Well-defined rhombic dodecahedron or cubic	Garnet Group Fe,Mg,Ca, Al silicates
	Green, yellowish green H = 6.5-7; D = 3.5-4.5 Granular glassy crystals	Olivine Mg_2SiO_4 to Fe_2SiO_4
	Variety of colors: clear, milky, rose, citrine, smoky, purple H = 7; D = 2.65 Well-developed hexagonal prismatic crystals Conchoidal fractures	Crystal Quartz SiO_2
	Microcrystalline or cryptocrystalline Light color variety: chert Dark color variety: flint Red color variety: jasper "Rainbow" color display: opal Color-banded variety: agate H = 7; D = 2.65	Quartz varieties: Agate, chert, flint, jasper, opal

[**TABLE 1.5** Nonmetallic minerals (softer than glass)]

Cleavage	Other Diagnostic Properties	Name Composition
Prominent	Colorless to white Perfect cubic cleavage H = 2-2.5, D = 2 Salty and soluble in water	Halite NaCl
	Colorless to white Perfect cleavage in 1, poor in 2 other directions H = 2, D = 2.3	Gypsum $CaSO_4*2H_2O$
	Colorless, white, or pale yellow Perfect cleavages in 3 directions at 75° H = 3, D = 2.7 Reacts with HCl (fizzes)	Calcite $CaCO_3$
	Colorless, violet, yellow, blue, green Good cleavage in 4 directions H=4, D = 3 Transparent to translucent	Fluorite CaF_2
	Green to white 1 direction cleavage, thin scales H = 1, D = 2.8 Soapy, pearly	Talc $Mg_3Si_4O_{10}(OH)_2$
	Colorless to pale brown Perfect cleavage in 1 direction, thin elastic sheets H = 2-3, D = 2.8	Muscovite K,Al silicate
	Brown to black colors Colorless to pale brown Perfect cleavage in 1 direction, thin elastic sheets H = 2-3, D = 2.8	Biotite KMgFeAl silicate
	Green to dark green 1 cleavage direction, scaly sheets H = 2–2.5, D = 2.5-3.5	Chlorite Mg, Fe, Al silicates
Absent	Red, reddish brown, earthy H = 1.5 and may be variable across the sample Red streak	Hematite (Fe_2O_3)
	White to red H = 1.2 Massive (irregular mass)	Kaolinate $Al_4Si_4O_{10}(OH)_8$

Mineral Identification Worksheet

Sample	Luster	Color	Streak	Hardness	Cleavage #	Cleavage Angle	Special Properties	Mineral Name	Mineral Group
1									
2									
3									
4									
5									
6									

Mineral Identification Worksheet

Sample	Luster	Color	Streak	Hardness	Cleavage #	Cleavage Angle	Special Properties	Mineral Name	Mineral Group
7									
8									
9									
10									
11									
12									

Supplementary Exercises for the Mineral Lab

Mineral resources are widespread in our daily lives, from the materials we use to construct our shelters, to the food we eat, and jewelry we wear. It is often interesting and educational to take a glimpse at how we use minerals to enrich our lives, either by necessity or for entertainment.

Birthstones

1. What is your birthstone?

2. What are the properties of your birthstone?

3. Do you like your birthstone? Do the properties say anything about your personality?

State Minerals

1. What is the Maryland state mineral?

2. What is the state mineral from one of your favorite places (other than Maryland) to visit, or a place you would like to visit?

Common Mineral Uses

Please list one example for each of the five common mineral uses in our daily lives:

Food _____

Construction _____

Jewelry _____

Make up _____

Industry _____

Adopt a Mineral

If you were given an opportunity to adopt a mineral from a museum collection, what would be your choice of the mineral? Explain your rationale.

LAB TWO

IGNEOUS ROCKS

LEARNING OBJECTIVES:

- Know the common important terms related to igneous rocks.
- Be able to understand and describe major mineral compositions and textures.
- Be able to identify igneous rocks and their major constituent minerals.
- Understand the processes by which igneous rocks are formed.

Overview

An igneous rock is a type of rock that forms from cooling and solidification of magma or lava.

The formation of igneous rocks represents the cooling history of earth and the earth's internal dynamics. Magma can cool and solidify either beneath the earth's surface or near and at the surface. The cooling process, the upward moving path and cooling rate of magma, dictates the characteristics of igneous rock textures, while the type of magma controls the composition.

Characteristics of Magma

The nature and characteristics of magma are controlled mostly by its silica content SiO_2 temperature, and the amount of dissolved gas. There are four general types of magma, classified by chemical composition. For the simplicity of this introductory-level class, the three common types are:

1. **Mafic (basaltic) magma:** often referred to as silica poor, average SiO_2 ~50%, high in Fe, Mg, Ca, low in K, Na

2. **Intermediate (andesitic) magma:** SiO_2 ~60%, intermediate in Fe, Mg, Ca, Na, K

3. **Felsic (rhyolitic) magma:** rich in SiO_2 ~70%, low in Fe, Mg, Ca, but high in K, Na

However, magmas can change in composition as they evolve over time:

Composition can change due to subtraction of materials—for example, crystal differentiation, such as olivine crystals removed from melt.

Composition can change due to additions of materials—for example, the assimilation from partial melting of country rocks.

Temperature of Magma

The temperature of magma varies with composition. Based on some laboratory and field observations, the general temperature ranges of the three common types are:

*   Mafic/basaltic magma: 1000 to 1200°C

*   Intermediate/andesitic magma: 800 to 1000°C

*   Felsic/rhyolitic magma: 650 to 800°C

Gases in Magma

Gases are typically dissolved in magma deep below the earth surface, but they form separate pockets of vapors as magma moves upward and pressure decreases, much like soft drinks or champagne. It is the dissolved gases that make volcanic eruptions explosive, because the volume of gas expands very quickly as pressure is reduced. Commonly occurring gases in magma are as follows:

- Most of the gases consist of water vapor (H_2O), carbon dioxide (CO_2), and sulfur dioxide (SO_2).

- The amount of gas in magma is also related to the chemical composition of the magma. Felsic (rhyolitic) type of magma usually has higher amounts of dissolved gas than mafic (basaltic) magma.

- In decreasing order of abundance, volcanoes release carbon monoxide (CO), hydrogen sulfide (H_2S), hydrogen chloride (HCl), hydrogen (H_2), methane (CH_2), and hydrogen fluoride (HF).

Viscosity of Magma

Viscosity is the resistance to flow of fluid matter. For example, honey is more viscous than water. The viscosity of magma is controlled by the silica content of the magma due to its chemical bonding and magma's temperature (see Table 2.1).

For example, basaltic magma has much lower viscosity (fairly fluid) than felsic magma, which still is millions of times more viscous than water. With an increase in temperature, the viscosity decreases. Imagine how a bar of solid chocolate melts and flows as it is heated up.

- Viscosity increases with increasing SiO_2 concentration in the magma; therefore, higher SiO_2 (silica) content magmas have higher viscosity than lower SiO_2 content magmas.

- Viscosity decreases with increasing temperature of the magma. Lower temperature magmas have higher viscosity than higher temperature magmas.

Apply What You Learned

1. Which type of magma has a higher viscosity? (felsic, intermediate, mafic)

2. Which type of magma commonly traps more gases? (felsic, mafic)

3. Is lava of high viscosity more explosive or less explosive?

TABLE 2.1 Magma composition and viscosity

SiO$_2$	TYPE	TEMPERATURE (centigrade)	VISCOSITY	GAS CONTENT	ERUPTION STYLE
~50%	mafic	~1200	low	low	commonly nonexplosive
~60%	intermediate	~1000	medium	medium	varied
~70%	felsic	~850	high	high	explosive

Magma Movement

Magma rises due to its thermal buoyancy (hot temperature) and high pressure in magma chambers, in contrast to its surrounding rocks. Magma has a lower density and moves upward. As it rises, it loses heat to the surrounding country rocks beneath the surface and to air and/or water as it approaches the earth surface. The cooling and solidification occur along the path(s) of magma's movement, resulting in different intrusive and extrusive structures.

Intrusive Structures

Intrusive structures are classified based on the size, shape, and relationships (concordant or discordant) with surrounding country rocks.

Batholith: the most massive discordant intrusion that is often associated with the roots of major mountain belts, with an exposed area greater than 100 square kilometers or 40 square miles

Stock: a smaller discordant intrusion than a batholith, with an exposed area between 1 and 100 square kilometers

Dike: a nearly vertical or diagonal discordant tabular intrusion, generally occurring along subsurface fractures

Sill: a more horizontal concordant tabular intrusion occuring between adjacent rocks layers

Laccolith: a small concordant intrusion in a mushroom-shaped cross section

Extrusive Structures

In volcanic eruptions, ejected volcanic materials often include lava, tephra (solid fragments: ash, dust, rock debris, volcanic bombs), and gases. The style or nature of eruption is determined by silica content, type of magma, and the amount of the dissolved gas trapped in lava. As a result, different extrusive structures are formed.

Shield volcanoes: typically a large, low gradient extrusive structure, resulting from massive mafic lava eruption

Composite volcanoes (*stratovolcano*): the most scenic cone-shaped volcano, resulting from the alternating building up of pyroclastic eruptions and intermediate and felsic lava

Lava fissure (*Pahoehoe* and *Aa*) *flow*: both formed from mafic lava flow; pahoehoe with a smooth glassy surface, Aa flow with a broken angular flow surface

Pillow lava: pillow-shaped structures with chilled volcanic glassy crust and fine-grained rock interior, resulting from the submarine eruption of mafic lava

---

Apply What You Learned

---

1. What is the volcanic structure of the Big Island of Hawaii? (composite volcano, shield volcano)

2. What is the volcanic structure of Mt. Rainier? (composite volcano, fissure flow, shield volcano)

3. Which type of lava flows faster? (Aa, Pahoehoe)

Textures of Igneous Rocks

Magma's movement and cooling history govern the texture of igneous rocks—that is, the size, shape, and arrangement of mineral particles. The main factor that determines the texture of an igneous rock is the cooling rate of magma. Other factors include heat diffusion rate, rate of crystal growth, and lava eruption style.

Generally speaking, minerals crystallize slowly and have large crystals when hot magma cools down slowly, otherwise rocks have fine-grained or glassy textures. If two stages of cooling occur—that is, a slow cooling process allowing for few large crystals to grow, followed by rapid cooling resulting in the growth of many smaller crystals—then the two-stage process results in a porphyritic texture. The following are common igneous rock textures:

- *Vitreous*: glassy, no crystals

- *Aphanitic*: fine-grained crystals, <2 mm, difficult to see

Igneous Rock Texture	Photo Characteristics
Extrusive — Aphanitic	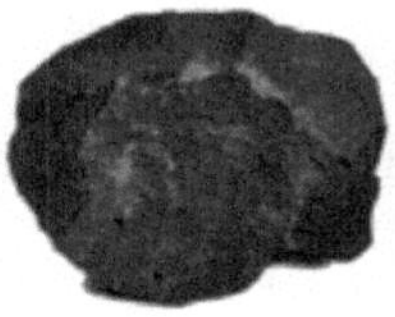 
Porphyritic-aphanitic	
Glassy	
Frothy, sponge glass	
Vesicular	
Pyroclastic	

FIGURE 2.1 Igneous extrusive rock texture (Credit: Mara Chen)

- *Phaneritic*: coarse-grained crystals, >2 mm, visible to naked eyes

- *Vesicular*: rich in cavities created by degasing, approximately 50% by volume

- *Porphyritic*: some large crystals (phenocrysts) in a finer matrix, suggesting a two-stage cooling process of the rock
  - *Porphyritic-phaneritic*: large crystals in a phaneritic matrix
  - *Porphyritic-aphanitic*: phenocrysts in an aphanitic matrix

- *Pegmatitic*: exceptionally coarse grained, >10 mm crystals, typically formed in late stages of crystallizing granitic magmas, water-rich slow cooling process

- *Pyroclastic*: characterized by a mixture of fragments, volcanic ashes welded together with angular rock fragments, volcanic glass, and perhaps minerals particles

Igneous Rock Texture		Photo Characteristics
Intrusive	Phaneritic	
	Porphyritic-phaneritic	

FIGURE 2.2 Igneous intrusive rock textures (Credit: Mara Chen)

Apply What You Learned

Using Figure 2.2 (above), answer the following questions:

1. Which has a coarser texture? (intrusive, or extrusive rocks)

2. What can you say about the crystal size in phaneritic texture?

3. What is the characteristic mineral size in an aphanitic texture?

4. Do all glassy textures look alike? What makes them different?

5. What is the main difference between a vesicular texture and frothy (sponge glass) pumice texture?

6. Why do extrusive rocks have a variety of textures?

Igneous Rock Mineral Composition

Igneous rock mineral composition is dictated by magma's chemical composition.

- **Felsic** igneous rocks are solidified from silica-rich magma and tend to be lighter in color (white, light gray, peach, pink, tan). Felsic is referred to as granitic.

- **Mafic** igneous rocks are crystallized and solidified from iron and magnesium-rich (and low silica) magmas and tend to be dark in color (black, dark brown, blackish gray, greenish black). Mafic is also referred to as basaltic.

- **Intermediate** igneous rocks are formed from intermediate magmas and tend to have shades or colors between felsic and mafic rocks (green, gray, brown). It is also referred to as andesitic in composition.

Apply What You Learned

---

1. Which type of igneous rocks has more varieties of minerals?

   (felsic, intermediate, mafic)

2. From felsic to intermediate to mafic, what is the trend of mineral assemblage?

   (more varieties of minerals, fewer minerals, just about the same)

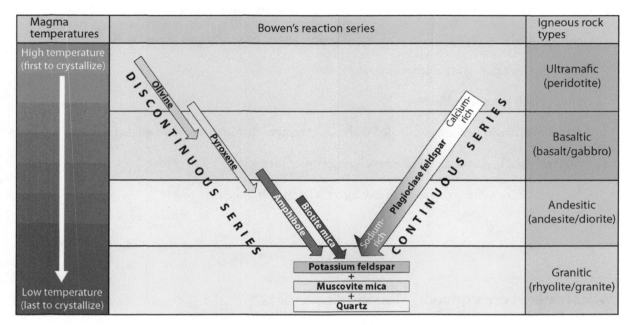

Magma temperatures	Bowen's reaction series	Igneous rock types
High temperature (first to crystallize)	DISCONTINUOUS SERIES — Olivine, Pyroxene, Amphibole, Biotite mica / CONTINUOUS SERIES — Plagioclase feldspar (Calcium-rich → Sodium-rich) → Potassium feldspar + Muscovite mica + Quartz	Ultramafic (peridotite)
		Basaltic (basalt/gabbro)
		Andesitic (andesite/diorite)
Low temperature (last to crystallize)		Granitic (rhyolite/granite)

FIGURE 2.3 Mineral crystallization sequence (Bowen's reaction series) and igneous rock composition (Source: USGS, http://www.nature.nps.gov/geology/education/images/GRAPHICS/KEMO-Bowens_series-01.jpg)

3. What are the represented rocks formed from mafic magma?

4. According to the Bowen reaction series in Figure 2.3, which type of igneous rock crystallizes first from a cooling magma? (felsic, intermediate, mafic)

5. What type of intrusive igneous rocks often has a salt-and-pepper appearance due to relatively equal abundance and mixing? (felsic, intermediate, mafic)

Common Minerals in Igneous Rocks

Although there are many different minerals that can be found in igneous rocks, only about 10 minerals are important to know to identify common igneous rocks. All the techniques and procedures you learned in mineral identification can be used; however, keep in mind that the minerals in the igneous rocks exhibit irregular geometric crystal shapes and are often much smaller in size, even microscopic.

- K-feldspar: pink with pearly luster, two-direction cleavages at nearly 90 degrees

- Quartz: colorless or smoky black, clear, no cleavage, conchoidal fracture, glassy luster

- Na-Plagioclase: a white feldspar with pearly luster and two cleavages at nearly 90 degrees

- Ca-Plagioclase: gray or dark gray, glassy or pearly luster, two cleavages at nearly 90 degrees

- Muscovite: clear or white mica, thin sheets

- Biotite: black mica, thin sheets

- Amphibole: black, slender, bar shaped with two cleavages that intersect at an angle far from 90 degrees

- Pyroxene: black, slab shape with two cleavages that intersect at 90 degrees

- Olivine: olive or yellowish green, glassy luster when fresh, no cleavage

Apply What You Learned

1. What minerals are commonly found in felsic rocks?

2. What minerals are commonly found in mafic rocks?

3. Please identify four minerals from the rock sample shown in Figure 2.4.

_____   _____   _____   _____.

Common Igneous Rocks Identification Procedure

Unlike minerals, rocks are not homogeneous, but aggregates of minerals; therefore, identifying rocks is normally harder than identifying minerals. For an introductory-level course like this one, however, you are expected to identify common rocks. Hopefully, the process is rather educational and interesting.

Igneous rock classification is based on two major aspects: rock texture and mineral composition.

Examining Rock Texture

Texture can be observed by the eye or by using a simple magnifying glass. Please hold the rock sample and rotate it under light and follow the flowchart (see Figure 2.6).

Identifying Mineral Composition

Unlike texture, the chemical composition of the rock cannot be directly measured or studied without instruments. Luckily, the colors of igneous rock are somewhat correlated to their chemical compositions. For example, iron and magnesium minerals are of darker colors, whereas K, Na, Al silicate minerals are of lighter colors. Therefore, colors can be used indirectly for figuring out an igneous rock's major compositions.

It is noteworthy to point out that there are exceptions to the color of an igneous rock due to crystal size, a minute amount of impurity, and weathering effects. For example, obsidian is typically of felsic composition, but it appears dark gray or black as a whole rock due to dense microscopic texture and a minute amount of iron content, but appears transparent or translucent in thin pieces or edges. Pumice is also volcanic glass, but may appear earthly dull due to weathering.

Overall, the color of a rock sample can be used as a general guide to get started. If you can see and identify the minerals, then you can look at the sample carefully and obtain more detailed information about the rock's mineral and chemical composition. If you can see a lot of quartz and/or K-feldspar in rocks, they are felsic. Igneous rocks with abundant olivine minerals are usually mafic. Igneous rocks with neither quartz nor olivine in them are likely intermediate. If you cannot see the minerals of the rock samples, then colors and special textures are often used as a guide. Common igneous rocks are shown in Figure 2.5.

FIGURE 2.5 Selected igneous rock samples (Credit: Mara Chen)

1: obsidian   2: pumice   3: scoria

4: basalt   5: rhyolite   6: porphyritic rhyolite

7: granite   8: pegmatite   9: syenite

10: diorite   11:gabbro   12: peridotite

1. Which type of igneous rock has a salt-and-pepper texture?

   (felsic, intermediate, mafic)

2. Generally speaking, which type of igneous rock has a light color?

   (felsic, intermediate, mafic)

After getting familiar with the identification of rock sample textures and compositions, you are ready to work on 12 igneous rock samples. Please follow the identification procedures discussed earlier and use Table 2.2 as a guide to complete the worksheet below.

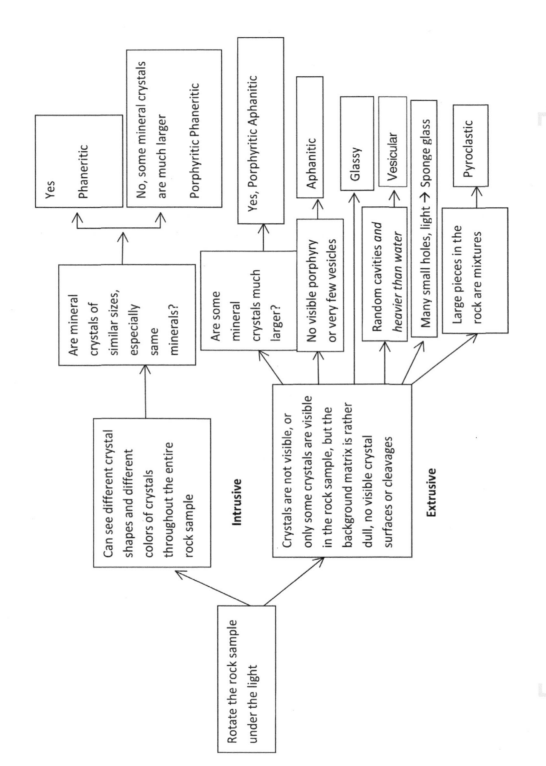

FIGURE 2.6 Flowchart of igneous rock texture identification (Credit: Mara Chen)

TABLE 2.2 Igneous rocks classification

Igneous Rocks		Felsic	Intermediate	Mafic	Ultramafic (Dark)
Intrusive Texture	Porphyritic-phaneritic	Porphyritic granite	Porphyritic diorite	Porphyritic gabbro	
	Phaneritic	Granite	Diorite	Gabbro	Peridotite
	Porphyritic-aphanitic	Porphyritic rhyolite	Porphyritic-andesite	Porphyritic basalt	
	Aphanitic	Rhyolite	Andesite	Basalt	Komatiite
Extrusive Texture	Vesicular	Vesicular rhyolite	Vesicular andesite	Vesicular basalt	
	Sponge glass, many small holes	Pumice			
	Solid glass, conchoidal fracture	Obsidian		Tachylyte	
	Pyroclastic	Tuff			
Common (mineral) composition		Quartz	Ca-Na Plagioclase	Ca-Plagioclase	Olivine
		K-Feldspar	Amphiboles	Olivine	Pyroxene
		Na-Plagioclase		Pyroxene	
		Mica			

Igneous Rocks Identification

Sample #	Texture	Overall Rock Color	Composition	Origin	Rock Name
1					
2					
3					
4					
5					
6					

Igneous Rocks Identification

Sample #	Texture	Overall Rock Color	Composition	Origin	Rock Name
7					
8					
9					
10					
11					
12					

Supplementary Exercises for the Igneous Rock Lab

1. Intrusive igneous rocks are known for their hardness, strength, and durability. In particular, granite also has a nice color and it has been used for exterior building stones, interior tables and countertops, and some of the world's famous monuments and memorials. Please use Google to check them out.

    Name a famous monument and type of igneous rock used.

    _____   _____

2. Walk around the university campus and note, is there any art statue made of igneous rocks? If so, where?

3. Does your home state have a state rock? If so, what is it? Granite is which state's official rock?

4. Igneous rocks were formed from cooling and solidification of magma. This process often led to higher concentrations of many valuable ores due to assimilating country rocks and fractional crystallization/magma differentiation.

    List two valuable ores that are commonly associated with igneous rocks.

    _____

    _____

LAB THREE

SEDIMENTARY ROCKS

LEARNING OBJECTIVES:

- Know the common important terms related to sedimentary rocks.
- Be able to understand and describe major textures.
- Become familiar with major sedimentary rock structures.
- Be able to identify common sedimentary rocks and the characteristics of texture and composition.
- Understand the major differences of two major subtypes, in terms of formation processes.

Nearly 75% of all rocks exposed on the earth surface are sedimentary, making it the most widely distributed rock on the planet's surface. The relative surface abundance is especially impressive, considering only 5% of the top 10 km of earth is made up of sedimentary rocks, while the rest (95%) are crystalline igneous, and metamorphic rocks.

Sedimentary rocks formed at or near the earth's surface and are composed of mainly weathered rock fragments and mineral particles, crystalline minerals from chemical solutions, and/or the remains of organic matter.

Sources of Sediments

Sedimentary rocks are made of sediments, which are products of different geological processes. Possible sources of sediments include:

- Physical weathering: bedrock may be physically broken down due to the changes of physical environment conditions, such as temperature and pressure

- Chemical weathering: the chemical solution and decomposition of bedrock

- Chemical precipitation: the formation of minerals from supersaturated mixtures of chemicals

- Organic accumulation: thick deposits of organic material accumulated over time

---

Apply What You Learned

Label the process for producing the sediments.

1. Rock fragments formed from repeated cycles of freezing and thawing _____

2. Rock salt formed from saline lake _____

3. Peat formed from dead plant matters in a swampy environment _____

Sedimentary Process

Based on the sources of sediments and subsequent rock-forming processes, sedimentary rocks are categorized into two subgroups: clastic and nonclastic rocks.

Clastic Sedimentary Rocks

Formation Process

The process is closely related to earth's denudation, including weathering, erosion, transportation, deposition, and lithification or diagenesis (compaction and cementation). Sediment grains or clasts, produced from physical and chemical weathering and erosion, are transported by running water, wind, and glacier ice, deposited in certain environments by these moving agents, and then chemically cemented into hardened rocks.

Weathering: the first step of earth's gradation (wearing down) process is breaking solid bedrock into smaller fragments or dissolving ions through chemical weathering processes

Erosion: the complex processes of attacking and removing rock fragments through mechanical scouring or chemical reactions by wind, water, and ice; overlaps with transportation

Transportation: the movement of sediments from one place to another by erosional agents, running water, wind, and glacier ice

Transportation process largely controls the size, rounding, sorting, and maturity of sediments. The longer the transportation time and distance from the original source area of the sediments, the better the rounding and sorting.

Size: sediment size ranges from huge boulders to clay mud particles. It is generally classified into gravel, sand, silt, and clay.

Rounding: Rocks break down and decompose into fragments. At the beginning, they tend to be angular. The transportation process tends to round off the sharp edges of sediments.

Sorting: the relative size range of sediments in a rock. Wind and running water are good sorting agents in transporting sediments, but glacier transport is not. If sediment grains are of similar size, then the rock is well sorted; otherwise, it is poorly sorted or somewhere in between.

Maturity: Sediment maturity is determined by texture and composition, reflecting the length of time the sediments have been in the sedimentary cycle. If sediments are well rounded and well sorted, then the rock is texturally mature. If sediments are of uniform composition and stable mineral particles, then the rock is compositionally mature.

Deposition: sediments are removed from suspension (transport) due to the reduction of energy of transporting agents. When a transportation agent no longer has sufficient energy to move its load, it will deposit the sediments. The depositional environments or locations represent the energy level and control the texture (grain size) and composition of the rock.

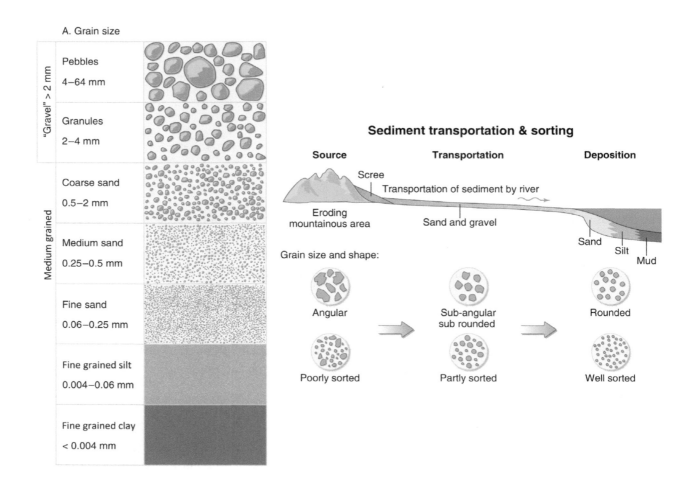

A. Grain size

"Gravel" > 2 mm	Pebbles 4–64 mm	
	Granules 2–4 mm	
Medium grained	Coarse sand 0.5–2 mm	
	Medium sand 0.25–0.5 mm	
	Fine sand 0.06–0.25 mm	
	Fine grained silt 0.004–0.06 mm	
	Fine grained clay < 0.004 mm	

Sediment transportation & sorting

Source — Transportation — Deposition

Scree
Transportation of sediment by river
Eroding mountainous area
Sand and gravel
Sand
Silt
Mud

Grain size and shape:

Angular — Sub-angular sub rounded — Rounded

Poorly sorted — Partly sorted — Well sorted

[**FIGURE 3.1** Texture, sorting, and rounding of clastic sediments]

Lithification: the rock-forming process in which the sediments are turned into hard rocks. It includes physical compaction and chemical cementation. Compaction reduces the pore spaces between particles and water content. Chemical cementation binds loose sediment particles together into solid rocks. Typically, water carries chemical compounds through the pore spaces of sediments, and those compounds gradually bind the loose particles together, forming sedimentary rocks. Three common cement are quartz, calcite, and iron oxides (commonly hematite).

Apply What You Learned

1. Does rounding of the sedimentary rock describe the shape of the rock sample or grain particles?

2. Which sample in Figure 3.1 exhibits good rounding? Which one shows angular sediments?

3. Are silt and/or clay particles rounded or angular?

4. Generally, coarse-grained rocks are ____ sorted. (well, poorly)

5. Fine-grained rocks are ____ sorted. (well, poorly)

6. Coarse-grained sedimentary rocks have better maturity. (true, false)

7. Which rock has a better compositional maturity? (quartz sandstone, arkose)

8. Name three possible depositional environments: _____, _____, _____

9. In a high-energy depositional environment, rocks tend to be poorly rounded. (true, false)

10. Does cementation control the physical and/or chemical properties of clastic rocks? Explain.

Nonclastic Sedimentary Rocks

Formation Process

Nonclastic rocks are also known as chemical and organic rocks. They are made of organic matter or pure chemical interlocked crystals, similar to igneous mineral crystals, except minerals grow from solutions, rather than from melts.

Precipitation: mineral particles formed from the precipitation from solutions or from chemical residues. Minerals grow from solutions as the chemical concentration increases when water evaporates.

Bioaccumulation of organic matter: After organisms die, their remains accumulate and are lithified into rocks.

Apply What You Learned

1. Give an example of rocks formed from chemical precipitation. _____

2. Give an example of rocks formed from accumulation of organic remains.
_____

Sedimentary Rock Texture

Clastic Rock Texture

The particle size, shape, and interrelationships determine the texture of rocks. A rock's texture reveals how it formed and the maturity of its composition/sediments. For clastic rocks, the texture is classified solely by the particle size, which is indirectly related to other aspects, such as sorting, rounding, and the maturity of sediments.

Coarse-grained (>2 mm) sediments are typically not sorted or matured. The shape can be rounded or angular or in between.

Medium-grained (1/16 to 2 mm) sediments typically went through a long-distance transportation and show a good degree of rounding and sorting.

Fine-grained (1/256 to 1/16 mm) sediments are barely visible, sorted, and mostly rounded. The rock of this texture feels mostly smooth to the touch, but gritty at fingertips or tongue tip.

Very fine-grained (<1/256 mm) sediments are clearly invisible grains, and the rock feels very smooth to the touch. Because of the very fine nature, rock of this type shows very fine layers or laminae.

Nonclastic Rock Texture

Crystalline: crystals visible to the naked eyes

Microcrystalline: very difficult to see without using a microscope

Oolitic: round nodules, varied in size; appear as concentric circles.

Fossiliferous/skeletal: abundant fossils

Note: Fossils can be found in both clastic and nonclastic rocks; the difference is relative abundance and overall rock composition.

Please complete Table 3.1.

TABLE 3.1 Common sedimentary rock texture   (Credit: Mara Chen)

#	Photo	Type (Clastic or Nonclastic)	Texture
1			
2			
3			
4			
5			
6			
7			

Select all that apply:

1. List rock samples in Table 3.1 that have good rounding, but poor sorting. _____

2. List samples in Table 3.1 that have good rounding and good sorting. _____

3. Are there any samples in Table 3.1 with poor rounding but good sorting? _____

Sedimentary Rock Classification

If rocks are made of clastic fragments, they are referred to as clastic sedimentary rocks. The further classification is based on the texture, the size of clastic particles. The maturity of the clastic sedimentary rocks is inversely related to the size.

If rocks are made of chemical and organic matter, they are known as nonclastic sedimentary rocks. The nonclastic rocks are further classified based on chemical composition, which is governed by the sedimentary environment in which they were formed. For example, coal is formed from transforming the dead remains of trees, ferns, and other plants.

Peat

Bituminous coal

Anthracite

FIGURE 3.2 Examples of organic rocks (Note: Anthracite is metamorphosed coal. Photo credit: Mara Chen)

Brief descriptions of common sedimentary rocks are presented in Table 3.2, and some common sedimentary rock samples are shown in Figure 3.3.

13: chert   14: siltstone   15: coquina

16: quartz conglomerate   17: red sandstone   18: argillaceous sandstone

19: shale   20: bituminous shale   21: limestone

22: dolomitic limestone   23: travertine   24: rock gypsum

FIGURE 3.3 Common sedimentary rock samples   (Credit: Mara Chen)

TABLE 3.2 Sedimentary rock classification

CLASTIC SEDIMENTARY ROCKS		
Texture	**Composition**	**Rock Name**
Coarse grained (Gravel >2 mm)	Rounded rock fragments of any types	Conglomerate
	Angular rock fragments of any types	Breccia
Medium grained (sand 1/16–2 mm)	Quartz and rock fragments	Quartz sandstone
	Quartz, feldspar minerals, and rock fragments	Arkose
Fine grained (silt 1/16–1/256 mm)	Quartz and clay minerals	Siltstone
Very fine grained (mud <1/256 mm)	Quartz and clay minerals	shale

NONCLASTIC SEDIMENTARY ROCKS		
Composition	**Texture Description**	**Rock Name**
Calcite (reacts with HCl)	Microcrystalline, conchoidal fracture	Micrite, microcrystalline limestone
	Crystalline	Crystalline limestone
	Banded in colors	Travertine
	Loosely cemented shell fragments	Coquina
	Abundant fossils: shells and other fragments	Fossiliferous limestone
	Microscopic shells and clays	Chalk
	Oolitic, spheric nodules	Oolitic limestone
Dolomite (only reacts with HCl in powered form)	Microcrystalline to crystalline	Dolomite limestone dolostone
Quartz (hard, glassy luster)	Microcrystalline crypocrystalline (finer crystal)	Chert (light color) Flint (dark color)
Gypsum (soft)	Fine to coarse crystalline	Rock gypsum
Halite (soft, salty)	Fine to coarse crystalline	Rock salt
Transformed plant matter	Very fine, luster shine, low density	Bituminous coal

Sedimentary Rock Identification Procedure

Rock identification is like doing detective work. You're provided with a tray of 12 samples and are expected to use your knowledge to figure out the properties and names for each sample.

Again, it is a process of identifying the physical/textual properties (particle size, shape, and arrangement) and chemical properties (composition and chemical reactions).

Please follow the flowchart Figure 3.4, and then complete the lab worksheet by using Table 3.2 as a reference.

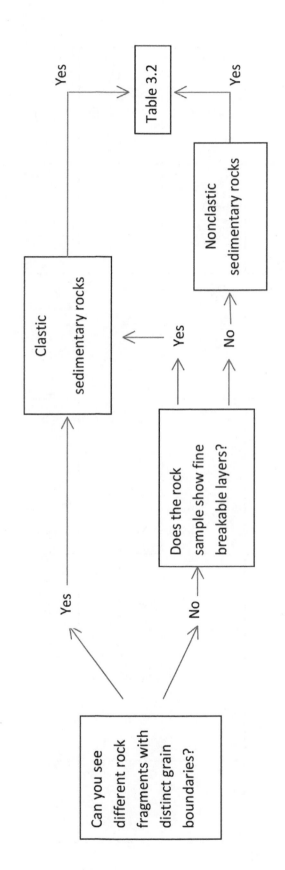

FIGURE 3.4 Flowchart for identifying sedimentary rocks (Credit: Mara Chen)

Sedimentary Rock Identification

Sample #	Subtype	Texture	Sorting/Rounding	Composition	Special Features	Rock Name
1						
2						
3						
4						
5						
6						

Sedimentary Rock Identification

Sample #	Subtype	Texture	Sorting/Rounding	Composition	Special Features	Rock Name
7						
8						
9						
10						
11						
12						

Sedimentary Rock Lab Supplementary Materials

1. Think about the important linkage between sedimentary rocks and natural resources, then describe the important uses of quartz sandstone, limestone, and rock salt.

2. What type of rocks were used in building the Getty Museum? What were the reasons behind the choice?

3. Are sedimentary rocks important to the exploration of energy resources? Explain.

4. The Grand Canyon provides us a spectacular glimpse into the profile of rocks, ranging in size by more than 18 miles in width at its widest span and more than 1 mile at its greatest depth. Most of the rocks on the canyon walls are sedimentary rocks, but how can you tell from a quick glimpse?

5. Many great national parks are in the southwest U.S. region. Please Google some of them and explain the aesthetical values of sedimentary rocks.

LAB FOUR

METAMORPHIC ROCKS

LEARNING OBJECTIVES:

- Know the common important terms related to metamorphic rocks.
- Be able to understand and describe the major factors that affect the composition and texture of metamorphic rocks.
- Be able to identify common metamorphic textures and major minerals.
- Be able to identify common metamorphic rocks.
- Develop a general understanding of metamorphism.

Definition

Metamorphic rocks are formed from the transformation and deformation of preexisting rocks in the solid state, without melting. They are closely associated with tectonic movement and naturally distributed along mountain belts and old continental shields.

Metamorphic rocks are widely used in construction and art for their prized texture patterns, colors, and durability. These properties reflect how they were formed.

Metamorphic Agents

Metamorphic rocks are changed, altered, or transformed preexisting rocks in a solid state. What made them change? As we have learned, rocks are aggregates of minerals. All minerals form and remain stable under certain environmental (physical and chemical) conditions. The changes in pressure, temperature, and chemical fluids trigger chemical reactions, which transform minerals when they are subjected to different equilibrium conditions. Consequently, new rocks may be formed from preexisting rocks. These new rocks are called metamorphic rocks and they commonly have different mineral assemblage and textures than the preexisting rocks from which they formed.

These environmental factors are referred to as metamorphic agents and include temperature, pressure, and active chemical fluids.

The physical factors, temperature and pressure, can change the size and shape of minerals, make them unstable, and promote chemical reactions to form new minerals. In addition, the rates and magnitudes of change often make metamorphic processes highly complicated. Among the three types of rocks, metamorphic rocks can have the most circuitous and dramatic stories to tell.

The chemical factor alters rocks' compositions and subsequently their appearance.

Temperature

In metamorphism, temperature ranges from above diagenesis (sedimentary environment) to partial melting (approximately 200°C to 800°C), before complete melting (igneous environment).

Average geothermal gradient: Temperature increases with depth in earth at a rate of about 25°C/km due to deep burial.

High geothermal gradient: Temperature can increase at a higher rate than the average geothermal gradient due to igneous intrusion.

Effects of temperature change:

- Break down preexisting minerals

- Speed up chemical reactions

- Promote partial melting and recrystallization

Pressure

The pressure of metamorphism varies a great deal across different environmental settings as well. It is estimated that the baseline pressure for metamorphism is about 300 MPa (mega pascals) or 3000 atmospheres of pressure. There are two types of pressures exerted on rocks.

Confining pressure: increases with depth of burial and exerts equally from all directions; known as hydrostatic stress or confining pressure

Differential pressure: increases due to tectonic movements and exerts unequal forces on the rocks in certain directions; known as differential or directional pressure/stress

Effects of pressure change:

- Pack particles closer

- Cause ductile/plastic deformation

- Create distinct alignment texture (foliation)

Active Chemical Fluids

Rock pore spaces often contain liquid, most often water. As a universal solvent, water can dissolve more substances than other liquids. Inside earth, water physically interacts and chemically reacts with surrounding rocks, exchanging ions and causing the alteration of existing minerals and formation of new minerals.

Effects include:

- Enhances migration of ions

- Aids in recrystallization of existing minerals

- Crystallizes new minerals, hydrothermally alters and recrystallizes existing minerals

Metamorphism

Metamorphism is the geological process that causes preexisting minerals to recrystallize, form new minerals, and causes new textures to develop. Common types of metamorphism include regional, contact, and hydrothermal, metamorphism.

Regional metamorphism: commonly associated with mountain building due to tectonic collision and subduction; involves a large region. Both temperature and pressure are important metamorphic agents.

Contact metamorphism: occurs in a relatively narrow contact aureole zone, surrounding large igneous intrusions due to high temperature. Sometimes, it is referred to as a baking process. Temperature is the dominant metamorphic agent.

Hydrothermal metamorphism: hot chemically reactive liquid is the key factor in altering preexisting rocks. It often results in forming many hydrothermal minerals and ore deposits.

Apply What You Learned

---

1. Can different types of metamorphism overlap/occur at the same area? (yes, no)

2. Can different types of metamorphism produce similar metamorphic rocks? (yes, no)

3. Where can you normally find metamorphic rocks?

Metamorphic Grade

Metamorphic rocks are products of increased temperature and pressure. Different environmental conditions produce different mineral assemblages. Metamorphic grade reflects the intensity of metamorphism. Some index minerals in the metamorphic rocks often serve as geothermometers and geobarometers, reflecting the metamorphic T/P conditions.

Common Minerals Associated with Metamorphic Rocks

Metamorphism intensity is often categorized into three levels: low, medium, and high metamorphic grade, based on the temperature and pressure range. Figure 4.1 shows the metamorphic grades and associated common minerals.

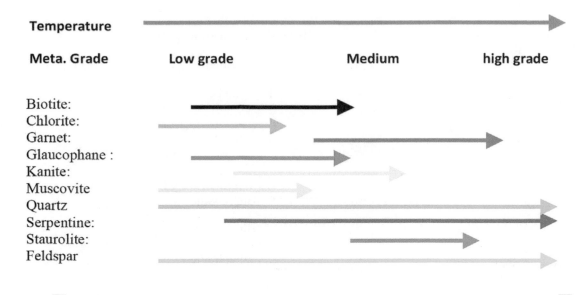

FIGURE 4.1 Metamorphic grades, the indicators of the intensity of metamorphism (Credit: Mara Chen)

Protolith

The protolith is the preexisting rock or parent rock that gets deformed and becomes a metamorphic rock. It can be an igneous, sedimentary, or metamorphic rock. Figuring out the protolith can be both interesting and challenging for many petrologists.

Of course, it is neither easy nor really a meaningful approach to examine the protolith of rock samples in the lab setting without a broader geologic context in the real world. Nonetheless, it is still interesting to make an educated guess theoretically, especially because the parent rocks' lineage for some metamorphic rocks is straightforward and simple to establish. Figure 4.2 represents some general metamorphism pathways for the selected metamorphic rocks.

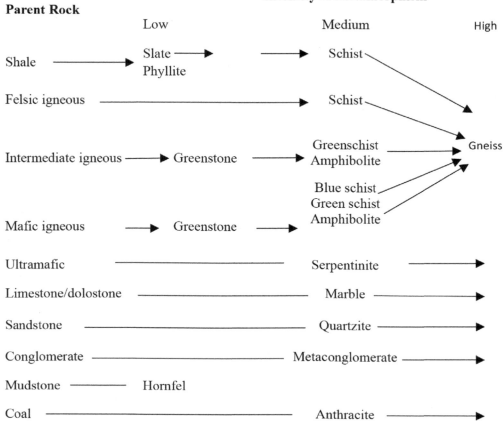

Parent Rock

Intensity of Metamorphism

Low | Medium | High

Shale → Slate → Phyllite → Schist

Felsic igneous → Schist → Gneiss

Intermediate igneous → Greenstone → Greenschist Amphibolite → Gneiss

Mafic igneous → Greenstone → Blue schist Green schist Amphibolite → Gneiss

Ultramafic → Serpentinite →

Limestone/dolostone → Marble →

Sandstone → Quartzite →

Conglomerate → Metaconglomerate →

Mudstone → Hornfel

Coal → Anthracite →

FIGURE 4.2 Possible pathways from parent rocks to transform into metamorphic rocks   (Credit: Mara Chen)

Apply What You Learned

1. What metamorphic rocks have a relatively simple pathway of transformation/ metamorphism?

2. Which metamorphic rock has very complicated pathways of transformation/meta morphism? Why?

3. Give two examples of low metamorphic grade rocks. _____, _____

4. Give two examples of high metamorphic grade rocks. _____, _____

Metamorphic Rock Texture

Metamorphic rocks formed under much higher/elevated temperature and pressure beneath the earth surface are made largely of mineral crystals. Again, the size, shape, and orientation of the mineral particles give metamorphic rocks different textures. In particular, the differential pressure gives some metamorphic rocks a unique texture, called foliation.

In general, metamorphic rocks have two major types of texture: foliated and nonfoliated.

Foliation or Foliated Texture

Foliation is the texture in which the platy or elongated mineral particles have preferred orientation or are aligned in a parallel or subparallel arrangement. As a result, foliated metamorphic rocks have flat layers or contorted layers or rock cleavage appearance. Because of different metamorphic grades, foliation is further classified into four different levels: slaty, phyllitic, schistosity, and gneissic banding.

Slaty cleavage: This texture results from the parallel orientation of microscopic mineral grains. In the field, it gives the rocks tendency to separate along parallel planes, which is known as slaty cleavage.

Phyllitic texture: Very similar to slaty cleavage, even often referred to as slaty cleavage, formed by the parallel arrangement of platy minerals, especially mica. However, the mineral particles are much larger than those in slate. So rocks of this texture have a satin or silky sheen luster due to the larger mica particles.

Schistosity or schistose texture: This results from the subparallel to parallel orientation of visible platy or elongated minerals such as chlorite, micas, and amphiboles. The foliation can be flat or crenulated, and rocks of this texture appear very shinny due to the crystal or cleavage surfaces of those platy minerals.

Gneissic banding: This is a high grade, coarsely foliated texture in metamorphic rock in which the minerals have been segregated into alternated dark-color and light-color bands. The light-colored layers are commonly made of nonplaty minerals like quartz and feldspars, whereas thinner dark bands are made of biotite and amphibole/hornblende minerals.

Nonfoliated Texture

Nonfoliated metamorphic rocks do not contain minerals with visibly preferred orientation. Nonfoliated rocks commonly contain so-called equal-dimensional mineral particles, often predominantly one type of mineral, like quartz, calcite, or dolomite.

Another nonfoliated texture is physical deformation, not preferred alignment of platy minerals. For example, conglomerate got metamorphosed and the pebbles are deformed and even aligned.

1. Please label each of the photos in Figure 4.3 with the appropriate texture name: slaty cleavage, schistosity, gneissic banding, or nonfoliated.

2. Is the size of mineral particles related to metamorphic grade?

3. Does a coarser-grained metamorphic rock have a higher or lower metamorphic grade?

4. Foliation is commonly related to _____ metamorphism. (regional, contact)

Metamorphic Rock Classification

In the field, metamorphic rocks are often classified by the three types of metamorphism: regional metamorphic rocks, contact metamorphic rocks, and hydrothermal metamorphic rocks.

Hand specimens are normally classified by their texture into foliated and nonfoliated rocks, and then by mineral compositions. Table 4.1 is a summary of common metamorphic rocks.

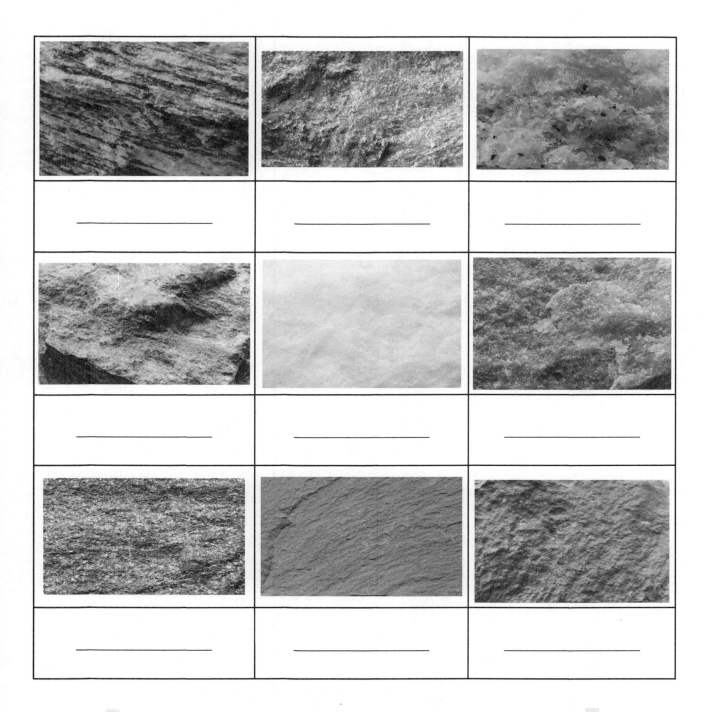

_____   _____   _____

_____   _____   _____

_____   _____   _____

TABLE 4.1 Classification of common metamorphic rocks

	Texture	Typical Composition	Characteristics	Rock	Metamorphic Grade	Possible Parent Rock
Foliated	Very fine (slaty)	Quartz, mica, chlorite	Rock cleavage, splits into flat pieces	Slate	Low	Shale
	Fine (phyllitic)	Chlorite, mica, quartz, feldspar	Fine grained, satin sheen	Phyllite	Low–medium	Shale, slate
	Medium coarse	Micas, chlorite, amphibole, feldspar	Visible minerals, distinct foliation	Schist	Medium–high	Slate, sedimentary and igneous rocks
	Medium coarse	Amphiboles, plagioclase	Rich in bar-shaped amphiboles	Amphibolite	Medium–high	Mafic igneous rocks
	Coarse (gneissic banding)	Quartz, feldspars, amphibole, micas	Distinct alternate dark and light mineral bands	Gneiss	High	Schist, or other rocks
Nonfoliated	Fine–medium	Chlorite, epidote amphibole	Green color, fine grained	Greenstone	Low–medium	Mafic igneous rocks
	Fine–medium–coarse	Calcite	Interlocking crystals, reacts with HCl	Marble	Low–high	Limestone or dolostone
	Medium–coarse	Quartz, rock fragments	Hard, dense	Quartzite	Medium–high	Quartz sandstone
	Coarse	Deformed fragments or pebbles	Flattened pebbles	Metaconglomerate	Medium–high	Conglomerate
	Lustrous	Carbon	Shiny, conchoidal fracture	Anthracite	High	Low-grade coals

Description of Common Metamorphic Rocks

Foliated and nonfoliated rocks with certain mineral assemblages are given specific names,. The following are common metamorphic rocks and their distinguishing characteristics (some of them are shown in Figure 4.4).

- *Slate:* fine-grained metamorphic rock, possessing slaty cleavage foliation; varied colors due to different composition

- *Phyllite:* similar to slate, a fine-grained, low-grade foliated metamorphic rock; however, shinier than slate due to larger mica mineral particles

- *Schist:* medium to coarse-grained foliated rocks, with varied colors and mineral compositions (Typically, color or important mineral[s] is[are] written into part of the metamorphic rock names [e.g., mica schist, greenschist, tremolite schist, garnet schist].)

- *Gneiss:* coarse-grained and banded foliation known as gneissic banding; the banding can be flat or crenulated

- *Amphibolite*: medium or coarse grained, dark colored rocks rich in elongated mineral hornblende and plagioclase; variable foliation texture with linear shiny bar-shaped minerals when they are aligned; can be nonfoliated when amphiboles are not aligned

- *Greenstone:* low grade metamorphic rock rich in chlorite, typically nonfoliated or weakly foliated; metamorphosed from mafic igneous rocks

- *Marbles:* rocks composed of calcite and less commonly of dolomite; may have pseudo-foliation due to impurity, such as micas

- *Quartzites:* stable over a wide range of pressures and temperatures; differs from quartz sandstone as it is made of interlocking crystals of quartz and has a sugary appearance due to the shattering of quartz crystals

- *Serpentinite:* rocks that consist mostly of serpentine minerals, formed from hydrothermal metamorphism of ultramafic igneous rocks

- *Skarns:* metamorphic rocks from contact metamorphism and hydrothermal metasomatism of carbonate rocks, but also rich in calcium and magnesium silicate minerals like andradite, epidote, diopside, and wollastonite

- *Anthracite:* lustrous shiny hard coal, metamorphosed from bituminous coal; dark shiny color, sometimes with an iridescence appearance and light in density.

25: gneiss	26: grantoid gneiss	27: schist
28: garnet schist	29: hornblende schist	30: quartzite
31: red slate	32: gray slate	33: white marble
34: pink marble	35: serpentinite	36: soapstone

Metamorphic Rock Identification

Like other major types of rocks, texture and composition are keys to metamorphic rock identification. It is always a little challenging to differentiate foliated from nonfoliated textures.

For this week's lab, you are provided with another tray of 12 rock samples and expected to identify them and complete the lab worksheets. Please identify texture first and then follow the flowchart in Figure 4.5 by examining the minerals in a rock sample to determine if a rock is foliated or nonfoliated. Then examine the minerals' physical and chemical properties in a rock, such as size, shape, color, hardness, chemical reactions, etc., and compare them with the rocks described in Table 4.1 to identify specific rock sample.

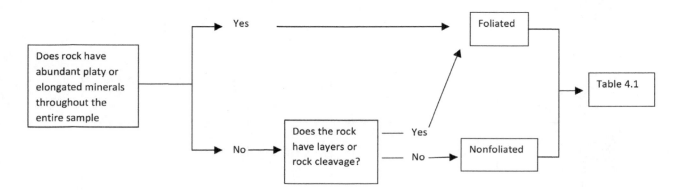

FIGURE 4.5 Identification flowchart of metamorphic rock (Credit: Mara Chen)

Metamorphic Rock Identification

Sample	Texture	Composition	Special Features	Rock Name	Parent Rock	Metamorphic Grade
1						
2						
3						
4						
5						
6						

Metamorphic Rock Identification

Sample	Texture	Composition	Special Features	Rock Name	Parent Rock	Metamorphic Grade
7						
8						
9						
10						
11						
12						

Metamorphic Rock Supplementary Lab Materials

1. What are the common uses of metamorphic rocks that you're aware of?

2. Take a walk around campus and complete the following table.

Location	Type of rocks	Remark
Front steps of Holloway Hall		
First floor of Henson Science Hall		
The green rock in the hallway of the Commons		
Lab tables		

3. What metamorphic rock is commonly used in art? Where can you find high-quality rock of this type in the United States?

4. What would you use metamorphic rock for?

LAB FIVE

GEOLOGIC TIME

LEARNING OBJECTIVES:

- Learn the relative order of geologic periods on the geologic timescale.
- Learn how to place a sequence of rocks in a relative order of events.
- Learn how geologists radiometrically date rock and mineral samples.
- Correlate rocks based on lithology and fossil content (time).

Introduction: The Geologic Time Scale

Most graphic displays of the geologic timescale are condensed representations of the tumultuous 4.56 billion years that has shaped Earth's surface (see Figure 5.1). As we have discussed in class, many of the divisions and subdivisions of the geologic timescale are delineated based on mass extinction events, which have been subsequently dated using sophisticated radiometric dating techniques and/or concurrent range zones of index fossil assemblages found within the rocks.

EON	ERA	PERIOD	MILLIONS OF YEARS AGO
Phanerozoic	Cenozoic	Quaternary	--- 1.6 --
		Tertiary	--- 66 --
	Mesozoic	Cretaceous	---138 --
		Jurassic	-- 205 --
		Triassic	-- 240 --
	Paleozoic	Permian	-- 290 --
		Pennsylvanian	--330 --
		Mississippian	-- 360 --
		Devonian	--410 --
		Silurian	-- 435 --
		Ordovician	-- 500 --
		Cambrian	-- 570 --
Proterozoic	Late Proterozoic Middle Proterozoic Early Proterozoic		--2500--
Archean	Late Archean Middle Archean Early Archean		-- 3800?--
Pre-Archean			

FIGURE 5.1 Geologic timescale showing the relative sequence of geologic eons, eras, and periods, and the associated radiometric ages in millions of years ago
(Image from USGS at http://pubs.usgs.gov/gip/fossils/numeric.html)

Part I: Relative Geologic Time

When geologists study rocks in the field, they seek to collect information regarding the rock type (igneous, metamorphic, or sedimentary), grain size, sedimentary structures, and fossil content within the sequence of rocks exposed at the earth's surface, among many other properties. A critical step in reconstructing the record of geologic events preserved in a sequence of rocks involves placing these rocks in a relative sequence of events as they were laid down on the earth's surface. This strategy is called **relative dating,** and it is generally the first step in reconciling time of any rock exposed anywhere. Relative geologic time is determined using the six principles discussed in detail in class and outlined below.

1. **Principle of Superposition**: In an undisturbed succession of rocks, the rocks at the bottom of the succession were deposited first (are older) and the rocks at the top of the succession were deposited last (are younger).

2. **Principle of Original Horizontality**: When sediments are deposited under the influence of gravity (in water, air, etc.), the layers will deposit one on top of the other in horizontal layers.

3. **Principle of Lateral Continuity**: When layers of sediment are deposited, the deposits will be infinitely continuous, unless the deposition grades laterally into a different rock type, or is affected by subsequent processes, such as erosion or faulting.

4. **Principle of Cross-Cutting Relationships**: Any fault or igneous intrusion that cuts across other layers of rocks are younger than the layers they cut across.

5. **Principle of Inclusions**: Rocks that contain inclusions or fragments of other rocks (i.e., all sedimentary rocks) will be younger than the inclusions that are found within that rock.

6. **Principle of Fossil Succession**: Extinct organisms are found throughout the rock record as fossils. Because different organisms appeared and disappeared throughout geologic time in a definite order through evolution and extinction, we can use the occurrence of fossil organisms to determine the ages of different rocks.

Apply What You Learned

Using Figure 5.2, place the sequences of rock into the appropriate order, using the processes of: "deposition," "erosion," "faulting," "folding," and "intrusion." Be sure you call out which layers (#) were affected by the process above.

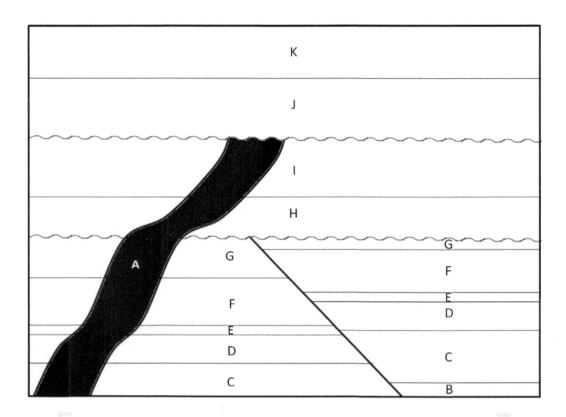

FIGURE 5.2 Relative dating exercise (Credit: Tom Cawthern)

Unconformities represent periods within a succession of rock in which layers have either been eroded or were never deposited to begin with. Simply stated, unconformities represent missing time within a stack of rocks. They are often identified in the rock record by using relative geologic time relationships and are more accurately dated using radiometric and/or the fossil content in adjacent (underlying and overlying) layers of rock. We have discussed three types of unconformities in class. These are:

1. **Disconformity**: an unconformity where flat-lying sedimentary rocks are overlain by flat-lying sedimentary rocks. Between these two rocks, layers have either been eroded or there was a period of no deposition. In either case, time is unaccounted for in the rock record. This generally follows the sequence of events: deposition (perhaps during high sea level) → erosion or nondeposition (as sea level falls) → deposition (as sea level rises again).

2. **Nonconformity**: an unconformity where sedimentary rocks are in contact with metamorphic or igneous rocks. Nonconformities may represent periods of significant erosion, for instance if a sedimentary rock is formed on top of an older igneous intrusion or a metamorphic rock layer, resulting in erosion of the older rocks. If this is the case, then inclusions of older igneous or metamorphic rocks would be expected in the younger sedimentary rock. Alternatively, nonconformities may form as a younger igneous intrusion, such as a dike, is injected into older sedimentary rocks. In this case, inclusions of the sedimentary rock may be found within the intrusion.

3. **Angular unconformity**: an unconformity formed by the general sequence of events: deposition of sedimentary rocks → folding of sedimentary rocks → erosion → deposition of new sedimentary rocks on top of the older, folded rocks. In general, angular unconformities form a sequence of rocks in which tilted or folded sedimentary rocks are found beneath flat-lying sedimentary rocks. Folding may occur as a result of mountain building events, called orogenies.

Youngest 7. _____

6. _____

5. _____

4. _____

3. _____

2. _____

Oldest 1. _____

Apply What You Learned

Identify the different types of unconformities present in the relative dating activity you completed using Figure 5.2. On this figure, be sure to circle and identify which feature represents the specific type of unconformity.

Part II: Radiometric Dating Techniques

Relative dating techniques are a useful means by which geologists are able to place layers of rock into a contextual framework that is both cost-effective and involves much less investment of time. Two very important dilemmas with relative dating techniques are that they are open to interpretation and that doesn't provide an actual age of the rocks. In other words, the relative order in which the rocks were created and altered since their formation is known using relative dating techniques, but the relative sequence of events is not founded in actual time.

In order to reconcile true geologic ages and in order to place relative time in the context of the geologic timescale, geologists utilize the fossil record as well as the natural radioactivity of elements contained within minerals from various rocks.

Some minerals, such as zircon, contain concentrations of naturally radioactive isotopes. These unstable radioactive isotopes decay by one half over a very precise time interval, known as the half-life (λ) (see Figure 5.3). Importantly, there are many different isotopic systems that geologist's can use to radiometrically date minerals and rocks. Some of these isotopic systems are best used for studying very old rocks, whereas others are best when studying very young rocks. When the radioactive isotope contained with a mineral decays, it produces a stable daughter isotope. The number of stable daughter isotopes increases through time as the number of radioactive parent isotopes decreases (see Table 5.1). Assuming that no parent or daughter isotopes are added or lost (i.e., the mineral is a closed system) throughout the course of time, the ratio of the parent-to-daughter isotopes can be used along with the half-life to determine the age of the mineral (and therefore, the rock).

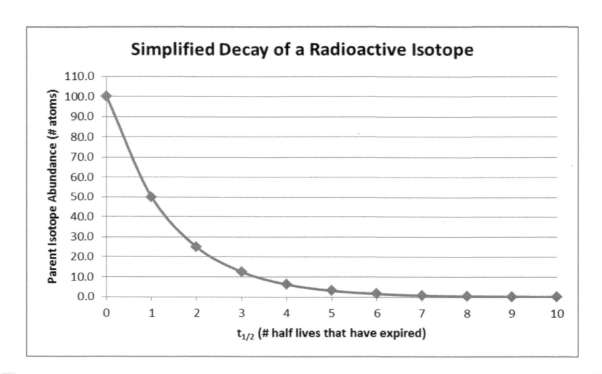

FIGURE 5.3 Simplified decay of a hypothetical radioactive isotope with a starting elemental abundance of 100 atoms. After one half-life, the number of remaining atoms of the radioactive parent isotope is 50; after two half-lives, the remaining number of radioactive parent isotopes is 25. As the number of parent isotopes diminishes by one half after each half-life, the number of stable daughter isotopes increases by the same amount as that which is lost by decay of the parent. (Credit: Tom Cawthern)

TABLE 5.1 Relationship between the number of parent and daughter isotopes and half-life (Credit: Tom Cawthern)

Parent	Daughter	$t_{1/2}$
100.0	0.0	0
50.0	50.0	1
25.0	75.0	2
12.5	87.5	3
6.3	93.8	4
3.1	96.9	5
1.6	98.4	6
0.8	99.2	7
0.4	99.6	8
0.2	99.8	9
0.1	99.9	10

* Note Age of rocks = # of HL * duration of the HL

Apply What You Learned

1. You have found an interesting rock outside your house that you want to have radio-metrically dated. You ship it off to the lab and the results are mailed back to you indicating that the parent-to-daughter ratio in the sample is 25:75. Moreover, the lab indicates that the isotopic system they used to ascertain this ratio has a half-life of 2,500,000 years. Knowing this information, determine the age of the sample you collected outside your house.

2. Using Figure 5.1, determine the geologic period when this rock was formed?

3. If the lab determined that the sample you discovered outside your home had a parent isotope composition of 12.5% and a known half-life of 500,000,000 years, determine the age of the sample in years. To which geologic time period does this rock belong?

Part III: Relative Time Dating: Lithologic and Biologic Correlations

Many geological field studies involve either mapping the spatial extent of rocks exposed at the surface, or studying a cross section of the vertical layering of rocks in a roadcut, river channel, or railroad cut. Often, these exposures are locally discontinuous and so the only way geologists can determine the true regional continuity of these layers is by correlating rocks across vast distances. In order to accomplish this monumental feat, geologists must employ the use of all of the relative dating techniques, but most importantly, the Principle of Lateral Continuity.

When the same geologic formation is locally discontinuous due to erosion, these layers can still be correlated over vast distances by using the unique characteristic features contained within each bed. For instance, if a sandstone layer contains coarse quartz grains, asymmetrical ripple marks, and fossil gastropods (snails) and bivalves (clams) in one location, and the same sandstone layer containing the same diagnostic features is encountered at a different location, then the layers can be correlated based on rock type. This technique is known as **lithologic correlation**.

Another correlation technique that is often employed is that of correlating based on geologic time. Many field studies focus on this method of correlation because it enables geologists to "see" the surface of the earth during a very specific window of geologic time. This allows geologists to recreate a map of ancient environments during a rather narrow period (a snapshot in time, really) relative to the entire geologic timescale.

In order to correlate rocks based on time, geologists must be able to first recognize time in the rocks. That might sound a little funny (How can anyone "see" time? Answer: geologists are really special people!), but it is a rather simple process. In order to determine the age of rocks in the field (and reconcile "time"), geologists study the fossils contained within sedimentary rocks. The Principle of Faunal (Fossil) Succession states that because different organisms appeared and disappeared throughout geologic time in a definite order through evolution and extinction, we can use the occurrence of fossil organisms to determine the ages of different rocks. Consequently, geologists must also be able to recognize individual species of particular fossil organisms—a field in its own right, called **paleontology**.

All extinct organisms have a **first occurrence** (the first time they appear in the rock record) and a **last occurrence** (the time when the organisms last appeared in the rock record). We can use this range zone to determine the potential age of a particular rock and if the same fossil species are found in rocks that are separated by vast distances, we can correlate these rocks based on the age of the rocks using the fossils, a process called **biologic correlation**. If a rock contains many different species that have different first occurrences and last occurrences, but those range zones all overlap, then we can better refine the age of the particular rock using the idea of **concurrent range zones** (see Figure 5.4).

Geologic Period	Age Range (mya)	Species A	Species B	Species C
Quaternary	1.6-0			
Tertiary	66-1.6			
Cretaceous	138-66			
Jurassic	205-138			
Triassic	205-240			
Permian	240-290			
Pennsylvanian	290-330			
Mississippian	330-360			
Devonian	360-410			
Silurian	410-435			
Ordovician	435-500			
Cambrian	500-570			

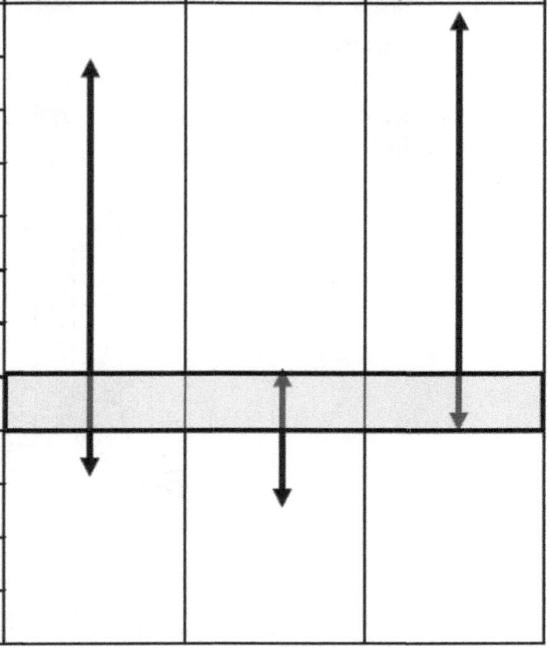

FIGURE 5.4 Concurrent range zone (gray box) of three fossil species found within a single rock layer. Each of the three fossil species have their very own first occurrence and last occurrence age. The overlapping age range is the age of the rock. (Credit: Tom Cawthern)

1. You have measured the characteristics of three different road outcrops that are separated from one another by different distances (see Figure 5.5). You have plotted the vertical variation in rock type and fossil content of each of these cross sections of rock. Using a pencil, correlate the rocks based on their lithology (lithologic correlation).

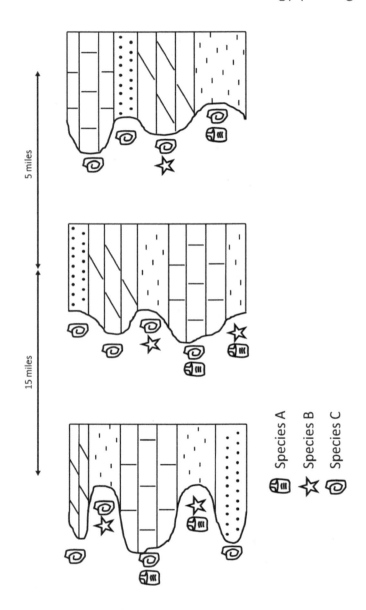

FIGURE 5.5 Measured sections from three different road outcrops (Credit: Tom Cawthern)

2. Now correlate the rocks based on their fossil content (biologic correlation).

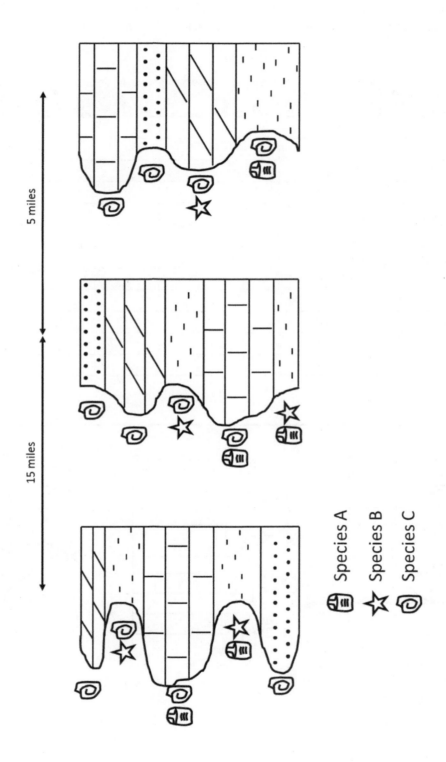

3. Assuming that Species A has a range zone of Cambrian-Ordovician, Species B a range zone of Ordovician-Mississippian, and Species C a range zone of Cambrian-Cretaceous, determine the geologic age range of each of the different layers in the three cross sections. Please use the same method shown in the example in Figure 5.4.

Geologic Period	Age Range (mya)	Species A 66-138	Species B	Species C
Quaternary	1.6-0			
Tertiary	1.6-6.6			
Cretaceous	138-66			
Jurassic	205-138			
Triassic	205-240			
Permian	240-290			
Pennsylvanian	290-330			
Mississippian	330-360			
Devonian	360-410			
Silurian	410-435			
Ordovician	435-500			
Cambrian	500-570			

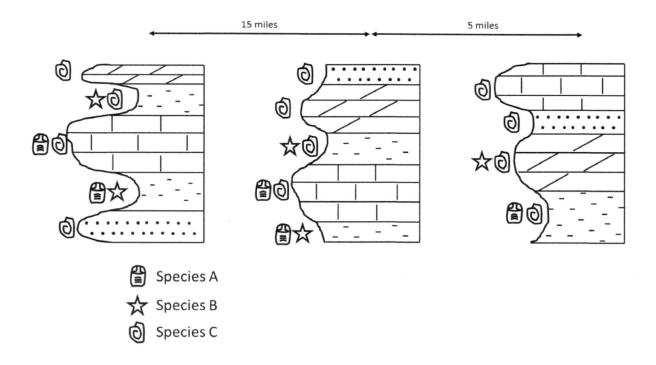

15 miles   5 miles

🔲 Species A

☆ Species B

◎ Species C

LAB SIX
TOPOGRAPHIC AND GEOLOGIC MAPS

LEARNING OBJECTIVES:

- Understand basic map elements: coordinates, symbols, scales, directions, legends.
- Learn how to read a topographic map: location, measurement, elevation, and other content.
- Learn how to construct a contour map.
- Learn how to construct a topographic profile.
- Understand basic elements of a geologic map (time scale, geologic formation, map legend).

Map Purposes, Geographic Grid, Map Elements

Maps serve many important purposes in our daily lives. They help us identify where we are, show what surrounds us, guide us where to go, informs us why things are where they are, and explain how the world has changed.

Different maps contain different information, but all maps are a graphic representation of information about a location on earth. Maps utilize visual language to tell a story about earth and the interactions between humans and the environment. The essential elements of a map that help users navigate an area include: the title, colors, lines, points, texts, legends, scale, direction, projection, map maker, and production date.

Geographic Grid: Longitude and Latitude

Maps cover specific locations on earth, and represent three-dimensional space in two dimensions of a piece of paper. Therefore, all maps need to use a coordinate system. The most commonly used coordinate system is the geographic grid. This system uses two important geographic reference lines and the center of the earth. One of the reference lines is the equator that divides the earth into two equal hemispheres: the northern hemisphere and the southern hemisphere. The lines of latitude or parallels (they are parallel to each other) measure the angular distances to the equator from north or south, ranging from -90°S to 90°N. The other is the prime meridian that passes through the Royal Observatory, Greenwich, London and divides the earth into two equal hemispheres: eastern and western hemispheres. The lines of longitude or meridians measure the angular distances to the prime meridian from east or west. The lines of longitude converge at the poles and have values ranging from 180°W to 180°E.

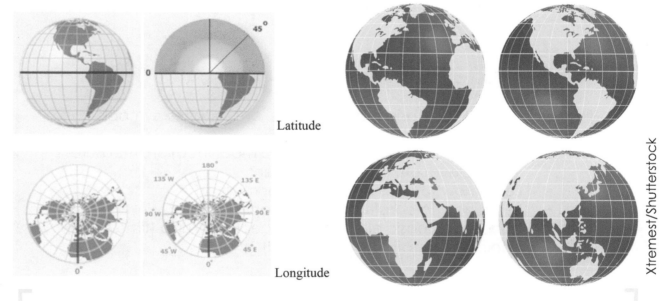

Latitude

Longitude

Xtremest/Shutterstock

FIGURE 6.1 Geographic grid: latitude and longitude. (Credit: nationalatlas.gov)

PHYSICAL GEOLOGY LAB EXPLORATION

Measurement

Any place in the world can be identified by using its longitude and latitude. To identify any place more precisely, the degrees of latitude and longitude can be further divided into minutes and seconds. For example

1° = 60' (one degree equals 60 minutes)

1' = 60" (one minute equals 60 seconds)

In addition to the geographic location, the ground distance of longitude and latitude can be measured as well.

Latitude:

1° = 69 miles (approximately)

1' = 1.15 miles

1" = 0.02 mile

Longitude:

A degree of longitude varies with latitude since they coverage toward the poles.

1° = 69 miles at 0° latitude, the equator

1° = 49 miles at 45° latitude

1° = 0 at the poles, 90° latitude

Apply What You Learned

1. What is the longitude and latitude of Salisbury, Maryland? _____

2. What is the longitude and latitude of your hometown? _____

3. What is the longitude and latitude of the opposite side of your hometown on earth?

_____

Map Elements

Maps tell stories, and map elements are a map's language, which helps users to better understand and read maps.

The essential map elements include scale, symbols, title, orientation, and legends.

Map Scale

Maps provide data and information. A map scale is vital for users to read and use the map correctly.

Scale representation: Scale can be represented as a ratio, a graphic bar, or a verbal sentence.

Ratio scale: a ratio between the map representation and real-world size; a common scale is 1/24,000

Graphic scale: a single line or bars with divisions, and each division represents a certain number of kilometers or miles

Verbal scale: uses a sentence to state the proportion of the map representation of the real-world features (e.g., "1 inch equals 10 miles" or "1 inch on the map equals 10 miles on the ground")

All three different forms of scales are interchangeable and can be converted from one to the other. In general, a small scale map is less detailed, but covers a larger area; a large scale map is more detailed, but covers a smaller area.

Topographic Map Symbols

Different maps utilize various symbols, and the USGS has a standard set of symbols (point, line, polygon shape, color, pattern, shade, lettering, etc.) to designate various features. More on this later in the lab.

Map Title

Title of a map can often say a lot about a map, such as location, theme, time, and so on.

Map Orientation

By mapmaking convention, the top of the map represents the geographic north (not magnetic north). However, maps can orient differently, as defined by a north arrow.

Map Legend

Although there are some common symbols, each map may have its specific symbols to convey information. A map legend shows map users what symbols mean in the map.

Types of Maps

There are many ways of rendering the representations for different types of information. There are two types of maps: general maps and thematic maps. General maps serve the general public and show information on locations, political boundaries, and physical landform features (water, city, forest, mountains, roads, etc.). Thematic maps show information on a specific topic/theme, such as climate, average precipitation, temperature, and geology.

To obtain skills in making and reading maps, students need background in different courses and years of experience. For this introductory-level physical geology course, students are asked to read and understand basic topographic and geological maps.

Topographic Map

A topographic map, also known as a topo or contour map, is a two-dimensional map that represents the physical configuration or the landscape of a place and portrays elevation variations using contour lines.

USGS Topo Map Symbols

USGS topo maps have standard map symbols, which are available online for download (http://egsc.usgs .gov/isb//pubs/booklets/symbolsnew/). Man-made cultural features are drawn in black; forests or other vegetation are in green; water features are in blue; contour lines are in brown; road networks are usually in red; features that have been updated using aerial photographs are usually in purple.

In addition to colors, features are shown in different forms of geometric elements such as point, line, and area. Line features can be straight or curved, solid or dotted, or a combination. The size of the lines shows the relative importance of the features.

Reading a Topo Map

Map symbols

Reading map elements, especially the graphic language, takes time. Understanding the colors, lines, and geometric shapes is key in map interpretation. As an example, answer the following questions regarding the area in and around Salisbury Quadrangle at: USGS http://store.usgs.gov/b2c_usgs/b2c/start/(xcm= r3standardpitrex_prd)/.do)

1. What is the name of the topo map?_____

2. Who published the map?_____

3. What is the publish date?_____

4. What is the latitude and longitude of the upper right corner of the map? _____

5. What is the fractional scale of the map?_____

6. What is the minute series map? _____

7. What does the green color stand for on the map?_____

8. What are the bold red lines? _____

9. What color represents contour lines? _____

10. What do the bold brown lines represent? _____

11. What do the black squares represent? _____

12. To which direction does the river flow? How do you know? _____ _____

Rules of Contour Lines

Contour lines are not the only features on a topo map, but are essential to fully understanding and appreciating the details on the map, and follow the set of rules below:

1. Contour lines link points of equal elevation on the earth's surface above or below a reference datum such as average sea level.

2. Contour lines of different elevations never intersect with each other, for different contours represent different elevations.

3. Contour lines never end in the middle of a map or run into a body of water.

4. The contour interval is the elevation difference between two consecutive contour lines.

5. Contour lines always bend upstream along a valley "V" and bend downslope along a ridge and appear straight at flat surfaces. The degree of bending corresponds to the depth of the valley or height of the ridge.

6. The spacing of contour lines indicate slope gradients. The more closely spaced the contour lines, the steeper the slope. The more widely spaced the contour lines, the gentler the slope or flat terrains such as plains and plateaus.

7. High elevation points such as mountains, hills, knobs, and pillars are shown as closed circular contours.

8. Low points (basins or craters) are shown as closed circular contour with hachure lines, indicating the elevation goes lower toward the center.

9. Index contours are darker brown lines that are labeled with elevations. Index contours help map readers determine the elevation at specific locations on the map quickly.

10. Contour lines are drawn by interpolating from elevation values at known points.

11. Local relief is the elevation difference between local high and low spots.

Apply What You Learned

Directions: Using the image below of the Menan Buttes Quadrangle as well as the whole quad online (USGS http://store.usgs.gov/b2c_usgs/b2c/start/(xcm=r3standardpitrex_prd)/.do), answer the following questions:

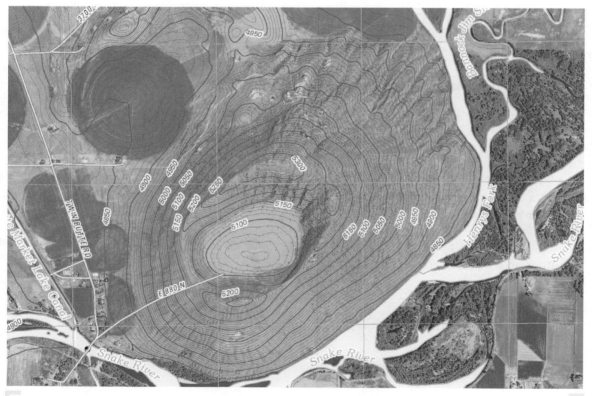

FIGURE 6.2 Part of the Menan Buttes Quadrangle
(Credit: USGS http://store.usgs.gov/b2c_usgs/b2c/start/(xcm=r3standardpitrex_prd)/.do)

1. What are the main geologic features represented by the contours with aerial photo backdrop?

2. How are the steep jagged walls represented by the contours?

3. What is the contour interval?

4. How could you tell whether there are changes or variations in the surface gradient?

5. Where is Menan Buttes located?

Construct a Contour Map

Procedures

The basic way we have discussed in class for constructing a contour map includes: [NT]

1. Find the highest point and lowest point.

2. Determine the number of contour lines based on the local relief from step 1.

3. Determine the first contour line using the contour intervals. For example, if you are asked to construct a contour map where the lowest elevation point is 8 and contour interval is 20, then the first contour would be 20, not 28.

4. Start with the lowest point and use the rule of interpolation to draw contours, until you finish all the contour lines determined from step 2.

5. Follow the "Rules of Contour Lines" presented earlier.

6. Please label each contour line as you finish drawing it.

1. Please complete this contour map using a 50-foot contour interval.

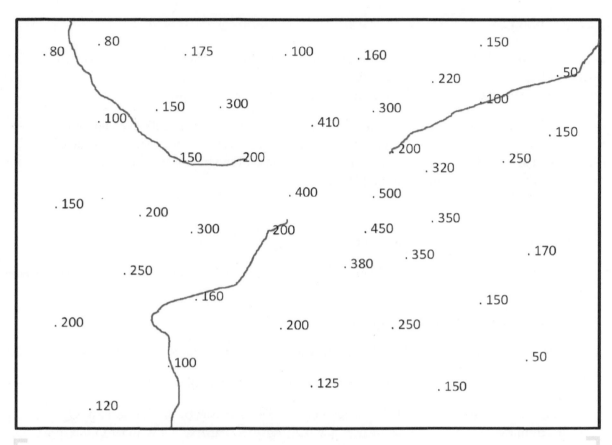

FIGURE 6.3 Elevation points for contour map. (Credit: Mara Chen)

2. Describe the landform represented by the contour map?

3. Where is the steeper slope?

4. How do contours bend when they cross streams?

Topographic Profile

Reading topo map symbols, interpreting contour patterns, and drawing contours are all important exercises in understanding topographic maps. In addition, constructing a topographic profile is another useful exercise for better understanding the landforms represented by topographic maps.

Topo maps are planar perspectives of the earth surface of a location from a bird's-eye view, while a topographic profile is a cross-sectional view, or side view, along a transect line at a certain elevation level, like what is revealed at a road cut. Beyond geological applications, a topographic profile is often important in understanding and solving civil engineering problems.

Procedure

- Place a blank piece of paper along the transect line.

- Mark down all the places along the blank piece of paper where the contour lines intersection the edge of the paper.

- Write down the elevation on each of these tic points.

- Transfer the points and their elevations to the right place on the profile (follow the lab instruction).

- Connect the points into a line, and you have your topographic profile done.

- Complete the topographic profile below (contour map source: part of the Castle Peak Quad, downloaded from the USGS at http://store.usgs.gov/b2c_usgs/b2c/start/(xcm=r3standardpitrex_prd)/.do).

Geological Map

A geologic map shows the distribution of the outcrop patterns of bedrocks, geologic structures, and resources in a geospatial context. It utilizes its own set of graphic language to represent the geologic features. The essential geological map elements include geologic time, contact lines (boundary lines between different geological units), symbols (colors, patterns, and text), scale, and legends. The standard geological symbols are available at this government site: http://ngmdb.usgs.gov/fgdc_gds/geolsymstd/download.php.

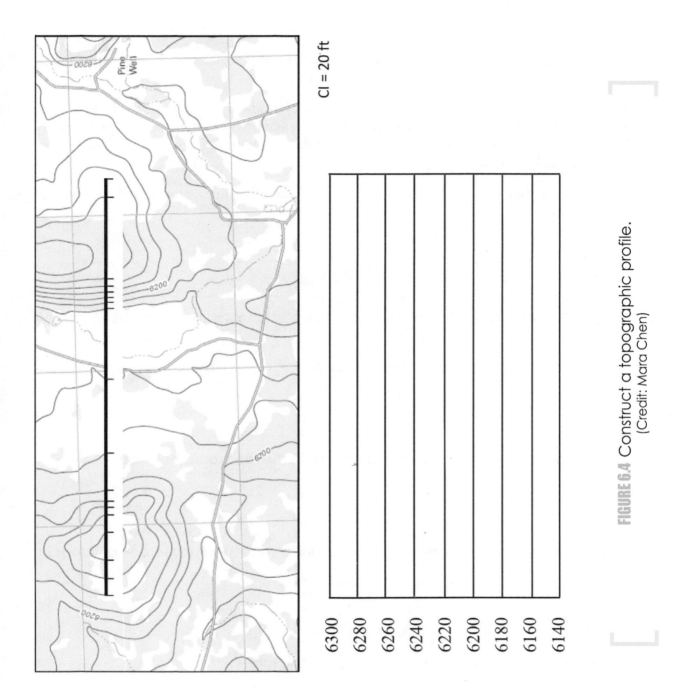

CI = 20 ft

6300
6280
6260
6240
6220
6200
6180
6160
6140

FIGURE 6.4 Construct a topographic profile.
(Credit: Mara Chen)

Geological Map Reading

The most striking feature of a geological map to a layperson is its colors. Each color represents a different geological unit formed at a different geological time.

It is important to understand the symbols of each map by studying the legend. Each geological legend box contains up to three important pieces of information: geological time, type of rocks, and the name of a geological unit. All the legend key boxes are arranged from the oldest to the youngest, from the bottom to the top.

Geological structures are faults, folds, domes, basins, and unconformities, represented by standard conventional symbols.

Apply What You Learned

Using the map of Maryland (Figure 5.5, or at https://jscholarship.library.jhu.edu/handle/1774.2/34143), answer the following questions:

1. What is the age of the oldest rock in Maryland?

2. What is the scale of the map? Please convert it to the ratio representation.

3. What is the age of the geologic units in Salisbury? Are there any bedrock outcrops in Salisbury?

4. Are there any faulting and folding structures on the map?

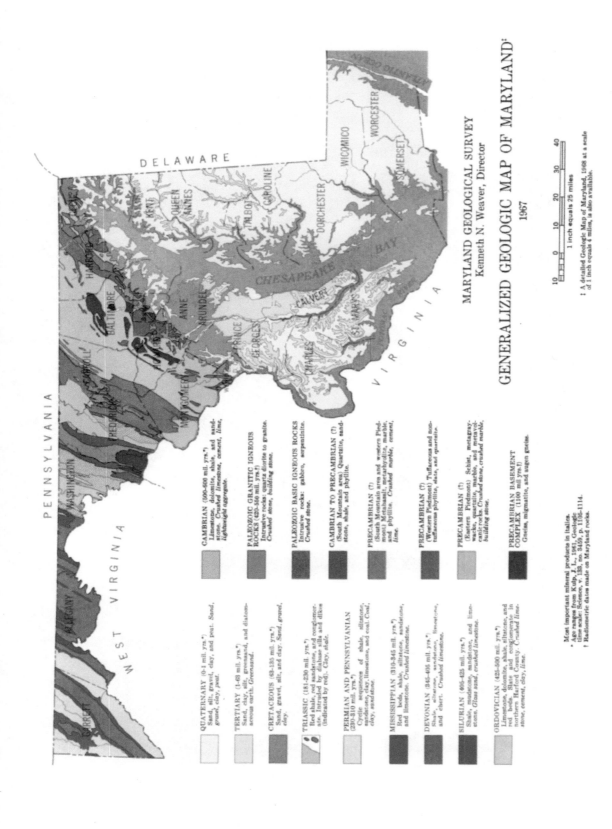

FIGURE 6.5 Generalized Geologic Map of Maryland.
(Credit: https://jscholarship.library.jhu.edu/handle/1774.2/34143)

LAB SEVEN

SURFACE WATER SYSTEM

LEARNING OBJECTIVES:

- Learn how to read and interpret stream hydrographs.
- Evaluate acceptable land use practices based on FEMA Flood Insurance Rate Maps.
- Evaluate the interaction of surface water flow and its influence on surface morphology.

Part I. Streams and Floods: Wicomico River/Beaverdam Creek near Salisbury, Maryland

In the first part of this lab we will use stream flow data from the South Prong of the Wicomico River/ Beaverdam Creek to:

• Understand how stream hydrographs work.

• Evaluate seasonal patterns in large flow rates.

Among its many responsibilities, the United States Geological Survey (USGS) maintains a network of over 7000 stream monitoring stations nationwide. The locations of these monitoring stations are situated along the banks of rivers throughout the United States. These monitoring stations serve to track the height of water flowing within the confines of a stream's channel continuously through the course of the year. Approximately 4000 of these locations have the capability to transmit their data via satellite link for "real-time" monitoring, which can be accessed for free via the USGS website. Each monitoring station is specifically calibrated to each stream such that a known reference level of water height corresponds to the point at which a stream overtops its banks. When the height of water flowing within a stream's channel becomes higher than the banks of the channel, the stream is said to have reached **flood stage,** and **flooding** will occur.

The southern reach of the Wicomico River in Salisbury, Maryland (recognized as Beaverdam Creek by the USGS), is monitored by the USGS at a "gaging station" located at the output of Schumaker Pond, where Beaglin Park Drive crosses Beaverdam Creek, near the Ward Museum of Wildfowl Art. Daily stream flows are reported in cubic feet per second (cfs) and are recorded along the Beaverdam Creek gaging station. You can access the flow data for Beaverdam Creek and at other stream gaging stations throughout Maryland at: http://waterdata.usgs.gov/md/nwis/rt.

Stream flow data are very useful for a variety of problems, such as:

• Providing data for forecasting and managing floods

• Setting minimum flow requirements for meeting aquatic life goals

• Estimating input rates of various pollutants into lakes, reservoirs, or estuaries

• Assisting recreational use (fishing, rafting, etc.)

• Assessing hydroelectric potential

Apply What You Learned

---

1. What happens if the water level exceeds the height of the stream banks?

2. Why does the USGS make such a concerted effort to maintain and continuously monitor streams throughout the United States?

3. What benefits could you foresee USGS stream monitoring stations being to you upon graduating from Salisbury University?

Evaluating Seasonal Patterns in Large Flow Rates

Streamflow data are usually reported in terms of **average daily values,** so there is a record for each day of the year. Flooding is associated with large flow rates, and it is customary to use the largest daily value for an entire year to study stream flooding. *The largest daily value for the entire year is called the Annual Peak Flow (APF), which may or may not exceed bank-full flow (i.e., flood).* Bank-full flow occurs when the water a stream is carrying exceeds the height of the stream's channel (or banks).

Streamflow data are generally reported in terms of *how much* water flows past the gaging station at a pre-determined unit of time (**discharge**). Streamflow discharge is therefore a volumetric rate of flow, and is determined by measuring the velocity of the moving water and multiplying that by the cross-sectional area of the stream channel.

$$\text{Discharge} = \text{Water Velocity (m/s or ft/s)} * \text{Cross-Sectional Area (m}^2 \text{ or ft}^2)$$

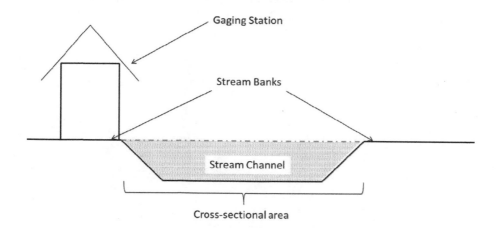

FIGURE 7.1 Cartoon showing the relative position of a stream gaging station situated along the banks of a stream. The stream is at bank-full, meaning that the height of the water is exactly at the height of the banks of the stream. Any addition of water to the stream channel will result in flooding.
(Credit: Tom Cawthern)

Discharge has been carefully calibrated at each USGS gaging station to correlate to the height of the water flowing past the monitoring station. Both stream gage height and discharge can be plotted with respect to time (days, months, years, etc.) to determine how the amount of water carried by each stream has fluctuated over the time interval studied. The histogram in Figure 7.2 shows the number of times the APF occurred in each month on Beaverdam Creek in Salisbury for a period between 1930 and 2012.

1. Note the two months in which APF occurs most often in Beaverdam Creek. Speculate why these months are more likely to experience the APF.

2. How many total years of record are there (HINT: the total number of years on record might differ from the total number of years that the stream hydrograph was constructed)?

3. What percentage of APFs occur in the months July, August, September, and October combined?

4. What are some possible causes of high water during this time of year?

5. What might have happened during the months of December, January, and February to cause APF events during the winter?

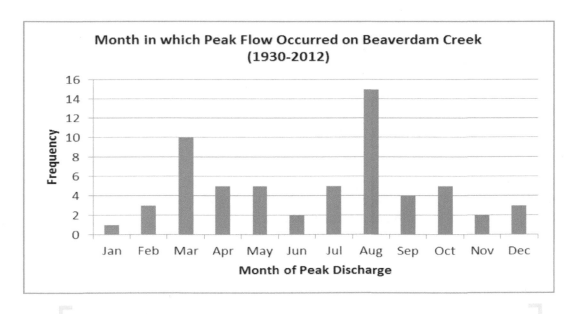

FIGURE 7.2 Annual Peak Flow (APF) recorded on Beaverdam Creek in Salisbury, Maryland, for the period 1930–2012.

PHYSICAL GEOLOGY LAB EXPLORATION

Evaluating Yearly Patterns in Large Flow Rates

Next, using the stream hydrograph (Figure 7.3) of Beaverdam Creek from August 2012 to July 2013, answer the following questions.

1. The daily mean discharge is plotted for the 12-month period. What is the peak discharge on this graph and when did it occur (month and approximate day)?

2. What is the lowest discharge during this 12-month interval?

3. How does the daily mean discharge for the time interval studied compare with the median daily statistics for the past 55 years?

4. What factors might account for the differences you observed in question 3?

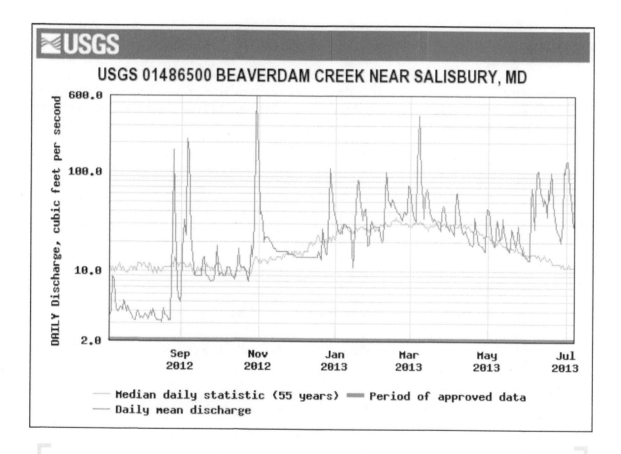

FIGURE 7.3 Hydrograph (discharge versus time) of Beaverdam Creek in Salisbury, Maryland, for the time period August 2012 to July 2013.

Part II. Flood Risk Map for Salisbury, Maryland

The Federal Emeregency Management Agency (FEMA) produces maps for the United States to help insurers determine flood insurance rates. A portion of a map that covers Beaverdam Creek in the vacinity of the Salisbury Zoo is shown on the last two pages of this lab. Note that this map went into effect in September 1984. More recent/updated FEMA maps for Maryland are available at: http://www.mdfloodmaps.net/dfirmimap/index.html.

1. Familiarize yourself with the map and the portion of the "Legend" provided. What do Zone A2 and Zone A3 refer to?

2. North and South Park Drive are found on either side of Beaverdam Creek. Based on the path of these roads and the FIRM map, would you say the roads were constructed before or after 1984? Why or why not?

3. The Salisbury Zoo is located on the southern side of Beaverdam Creek in between North and South Park Drive and Beaverdam Drive on the West and Memorial Plaza to the East. Locate the site of the zoo. Is the Salisbury Zoo safe from flooding?

4. What are some benefits of building on a 100-year floodplain?

5. What actions could be taken to reduce the probability of flooding?

Part III. Surficial Characteristics of Streams

The overland flow of water, generally referred to as surface water **runoff**, will become channelized and take on a characteristic pattern when viewed from above on an aerial photograph, on a satellite image, or on a topographic map. The drainage pattern of particular areas can be characterized as dendritic, radial, rectangular, or trellis in shape (see Figure 7.4).

- **Dendritic**: the most common drainage pattern morphology, forming a branching pattern similar to the roots of a tree or plant. Typically develops in regions underlain by homogeneous stratigraphy. Tributaries entering the main river channel intersect at angles <90 degrees.

- **Radial**: develops from surface water runoff flowing away from a central high point in all directions. Generally looks like the spokes of a tire.

- **Rectangular**: develops in regions where the underlying stratigraphy has been faulted. Surface water runoff will follow the fault traces at the surface. Subsequent displacement of a fault zone may result in the offset of a tributary or channel. Tributary streams typically merge at high angles (~90 degrees), but unlike trellis drainage patterns, the tributaries may be of much greater length.

- **Trellis**: typically develops above folded sedimentary strata (i.e., Appalachian Mountains). Short tributaries draining ridges enter the main river channel in the intervening valley at nearly right angles.

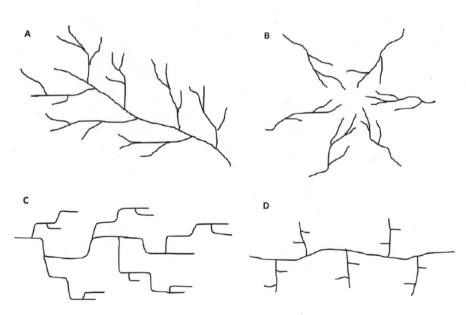

FIGURE 7.4 Surface water runoff patterns. A) dendritic drainage; B) radial drainage; C) rectangular drainage; D) trellis drainage. (Figure credit: Tom Cawthern).

Match the surface water drainage patterns on the topographic maps in Figure 7.5 with the correct terms, as defined above (dendritic, radial, rectangular, and trellis).

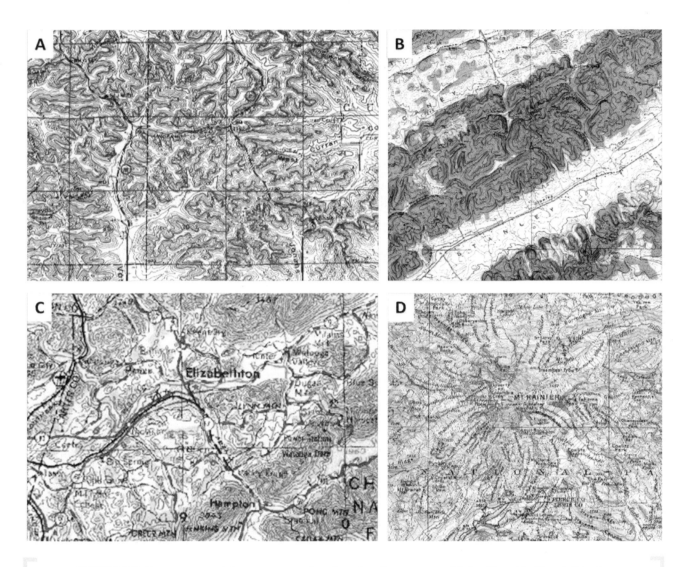

FIGURE 7.5 Topographic maps of selected areas within the United States, exhibiting each of the different surface water drainage patterns.
(Maps courtesy of the USGS)

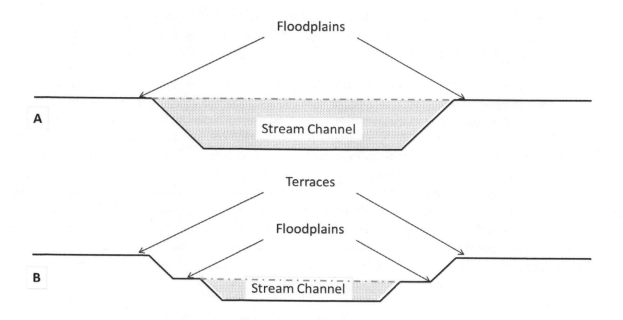

FIGURE 7.6 Cartoon illustrating the formation of terraces (abandoned floodplains) as a result of downcutting (Credit: Tom Cawthern)

Although the pattern of surface water runoff can be influenced by the rocks over which they flow, the shape of the ground surface may also be modified by this flowing water. Rivers and streams erode into surrounding rock and sediments in three dimensions: They carve vertically downwards (**downcutting**), they erode laterally (**lateral erosion**), and they erode parallel to the length of the stream through **headward erosion**. The process of downcutting (see Figure 7.6) results in the formation of **terraces**—that is, floodplains that have been abandoned due to the vertical carving of the stream's channel.

As water flows down gradient (from high elevation toward low elevation), they will expend most of their energy in places where the surface topography is steep. **Surface gradients** (the steepness of the ground surface) are greatest at the head of the stream and less at the mouth of the stream. Stream gradient is calculated by dividing the elevation change of the stream between two points over the distance between those two points. Both of these values may be determined from a topographic map: Elevation is determined from the contour lines, whereas the horizontal distance can be determined by using the scale of the map. Gradient is often expressed in ft. per mile, or meters per kilometer units.

$$\text{Gradient} = \frac{\text{Change in elevation between 2 points}}{\text{Horizontal flowing distance between those 2 points}}$$

The **head** of the stream is the portion that is located at higher elevation, closer to where the stream originates (i.e., mountains). By contrast, the **mouth** of the stream is where the stream gradient is lowest, generally closest to sea level. Where a stream flows into a large body of water, such as the ocean, a pile of sediment will be deposited. This pile of sediment is called a **delta**. Deltas may also form at the base of mountains in arid or semiarid regions with large topographic relief (i.e., steep mountains located adjacent to low basins) that experience sporadic, high rainfall. These deposits are called **alluvial fans**.

As water within a stream's channel migrates from high elevation down to low elevation, the water closest to the banks and the streambed moves slower than the water located closer to the surface and in the middle of the channel. This is because of boundary conditions imparted due to frictional drag along the stream channels. Most stream channels do not follow a purely linear (straight) course from their headwaters to their mouth. Instead, the channels bend (or **meander**) across the ground surface. Meandering streams cause the high energy "core" of the stream to deflect toward the outside bend of the meander. Consequently, erosion occurs on the outer bend of meanders, giving rise to the term **cutbank**. In contrast, much lower energies are associated with the inside bend of a meander, giving rise to the term **pointbar**. Here, sedimentation occurs, resulting in the addition of material. Through time, meanders will migrate across the ground surface as cutbanks erode and pointbars build up sediments. As a result, some meander bends may become **cutoff** from the migrating channel, forming an **oxbow lake**, a crescent-shaped body of water that represents the old stream channel. See Figure 7.7.

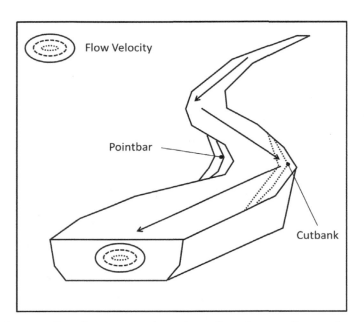

FIGURE 7.7 Schematic diagram showing the formation of pointbar deposits and cutbanks as a result of a meandering stream. Flow velocities are indicated by the concentric circles drawn above with higher flow regimes characterized by finer dashed lines. Arrows indicate the general sense of flow of the water within the stream channel. (Credit: Tom Cawthern)

PHYSICAL GEOLOGY LAB EXPLORATION

Use the maps provided in the classroom to answer the following questions.

Tenmile Point, Utah

1. What type(s) of drainage pattern do you see? (see Figure 7.4)

2. What evidence on this map indicates that the Green River has downcut into the under-lying rocks? Which side of the river bank has downcutting been more concentrated?

Devil's Elbow, Missouri

1. Based on its current location, explain why you think the town of Devil's Elbow is relatively safe from erosion.

2. Calculate the gradient of the Big Piney River (main river) AND the gradient of the tributary entering the Big Piney River across from Devil's Elbow. How do these gradients compare?

3. Explain why the contour lines are so close together along the eastern bank of the Big Piney River.

Death Valley, California

1. What happens to surface water runoff from higher elevation upon entering the valley below? What is the relationship between this pattern and the occurrence of sand dunes?

2. Explain why contour lines become increasingly more spread out the farther you get from the mountains on the western side of the map.

3. What is the surface gradient of Death Valley Junction in the north central corner of the map?

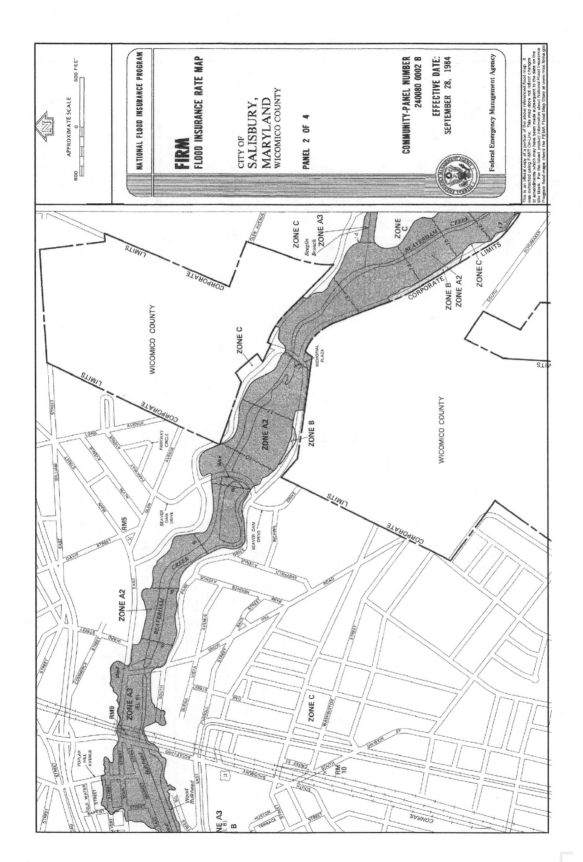

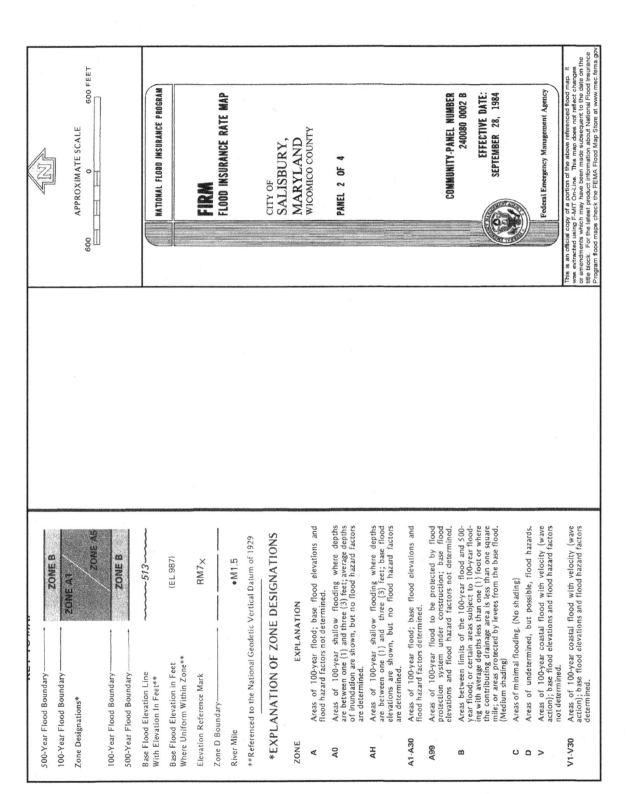

APPROXIMATE SCALE

600   0   600 FEET

NATIONAL FLOOD INSURANCE PROGRAM

FIRM
FLOOD INSURANCE RATE MAP

CITY OF
SALISBURY,
MARYLAND
WICOMICO COUNTY

PANEL 2 OF 4

COMMUNITY-PANEL NUMBER
240080 0002 B

EFFECTIVE DATE:
SEPTEMBER 28, 1984

Federal Emergency Management Agency

500-Year Flood Boundary

100-Year Flood Boundary

Zone Designations*   ZONE B

                                       ZONE A1
                                       ZONE A5

100-Year Flood Boundary   ZONE B

500-Year Flood Boundary

Base Flood Elevation Line   ——513——
With Elevation In Feet**

Base Flood Elevation in Feet   (EL 987)
Where Uniform Within Zone**

Elevation Reference Mark   RM7x

Zone D Boundary

River Mile   •M1.5

**Referenced to the National Geodetic Vertical Datum of 1929

*EXPLANATION OF ZONE DESIGNATIONS

ZONE	EXPLANATION
A	Areas of 100-year flood; base flood elevations and flood hazard factors not determined.
AO	Areas of 100-year shallow flooding where depths are between one (1) and three (3) feet; average depths of inundation are shown, but no flood hazard factors are determined.
AH	Areas of 100-year shallow flooding where depths are between one (1) and three (3) feet; base flood elevations are shown, but no flood hazard factors are determined.
A1-A30	Areas of 100-year flood; base flood elevations and flood hazard factors determined.
A99	Areas of 100-year flood to be protected by flood protection system under construction; base flood elevations and flood hazard factors not determined.
B	Areas between limits of the 100-year flood and 500-year flood; or certain areas subject to 100-year flooding with average depths less than one (1) foot or where the contributing drainage area is less than one square mile; or areas protected by levees from the base flood. (Medium shading)
C	Areas of minimal flooding. (No shading)
D	Areas of undetermined, but possible, flood hazards.
V	Areas of 100-year coastal flood with velocity (wave action); base flood elevations and flood hazard factors not determined.
V1-V30	Areas of 100-year coastal flood with velocity (wave action); base flood elevations and flood hazard factors determined.

LAB EIGHT

STRUCTURAL GEOLOGY

LEARNING OBJECTIVES:

- Learn to identify basic geological structures on a geological map.
- Learn how to read geological maps: map keys, time sequences, geological formation symbols, contact lines, contour lines, geological structures, and the attitudes of rock units (strikes and dips)
- Learn how to read geological profile or cross sections
- Understand directions and map scales
- Understand basic elevation change and local relief

Minerals and rocks provide the material foundation for earth landform features. The surface landforms are external manifestations of deformed and metamorphosed rocks and minerals. This lab focuses on how rocks deform to form different geological structures in response to different types of crustal stress. Through the lab, you would gain fundamental understanding and training in reading geological structural diagrams and geological maps.

Part I: Michigan. Answer the Following Questions Based on Figure 8.1:

1. What is the geologic structure of Lower Michigan?

2. Explain your rationale behind your conclusion.

3. What is the possible age of the structure, i.e., when was it formed?

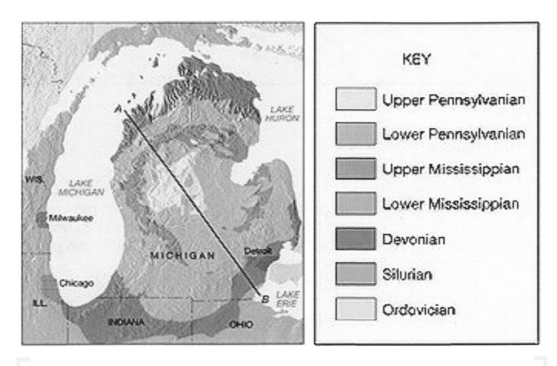

FIGURE 8.1 Geologic map of the Lower Michigan (Source: USGS)

Part II: Black Hill. Answer the Following Questions Based on Figure 8.2:

4. What is the structure of the Black Hill based on the map?

5. Explain your rationale behind your decision. What is the age of the oldest rock?

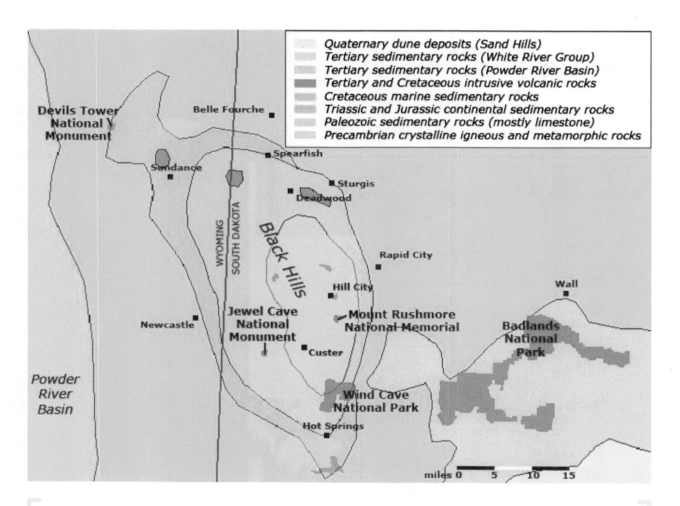

Quaternary dune deposits (Sand Hills)
Tertiary sedimentary rocks (White River Group)
Tertiary sedimentary rocks (Powder River Basin)
Tertiary and Cretaceous intrusive volcanic rocks
Cretaceous marine sedimentary rocks
Triassic and Jurassic continental sedimentary rocks
Paleozoic sedimentary rocks (mostly limestone)
Precambrian crystalline igneous and metamorphic rocks

FIGURE 8.2 Geologic Map of the Black Hills region in South Dakota
(Source: USGS)

Part III: Virtual Geological Journey of the Middletown Quad, Virginia

Reading geological maps, like reading road maps or climate maps, requires readers to understand map keys/legends, scale, orientation, title, colors and text symbols etc.

The Geological Map of the Middletown Quad, Virginia is available in the lab room. You may also find it the digital version of this map useful: http://ngmdb.usgs.gov/Prodesc/proddesc_19293.htm

Understand the Geological Map Elements:

6. In Om symbol, what does the uppercase letter 'O' denote Ω? How about small letter 'm'?

   Om   Martinsburg Formation (Upper and Middle Ordovician)

7. What is the geological term for the boundary line between two different rock units?

8. Please draw the following structure symbols:

   Thrust fault   _____

   Anticline   _____

   Syncline   _____

   Strike & dip   _____   (label 'strike' and 'dip' on the lines)

   Sinkholes   _____

   Quarry   _____

9. What is the map scale? _____

Virtual Geological Tour: Site 1 to Site 5 (All Diagrams Are Enlarged Portions of the Quad Map, Which Helps you to Find the Right Locations on the Map).

Site 1: Reliance

Begin your journey near the town of Reliance in the southeastern part of the quadrangle map.

What is your latitude and Longitude at the lower right corner of the quadrangle at this location?

_____ _____

What age are these rocks? _____

What type of rock is this? _____

What geologic structure can you observe using the symbols across the map area ? _____

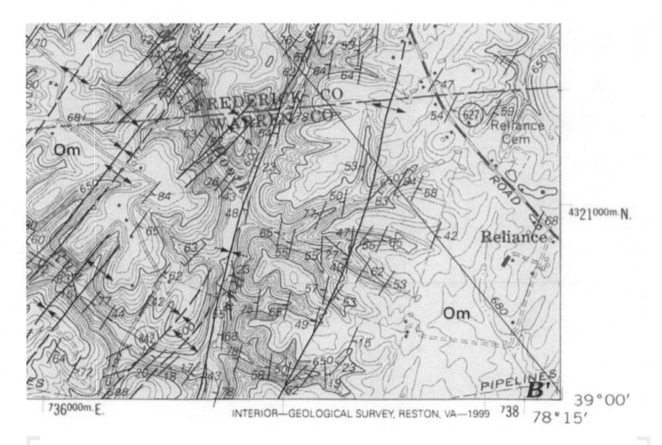

(Source: USGS)

Site 2: Cedar Creek.

What is the youngest rock formation you can observe in/around Cedar Creek? _____

How is this formation distributed relative to the Creek - near or far away? _____

When you look at the bottom of the map, you notice the scale of the map. Please express the scale in a sentence. _____

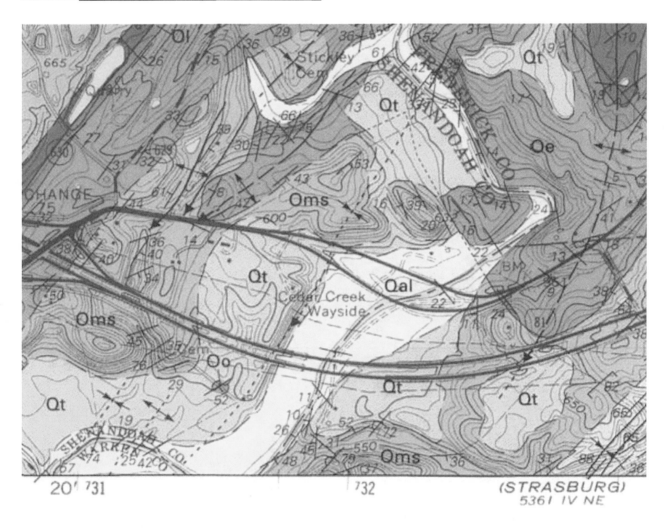

(STRASBURG)
5361 IV NE

SCALE 1:24 000

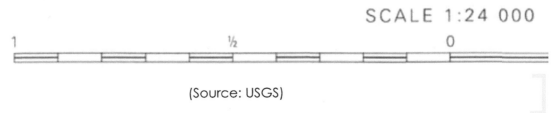

(Source: USGS)

Site 3: Oranda

Do you see any evidence of quarrying activity near the town of Oranda? What rock material is being quarried and processed here based on the map legend? _____

What are the geologic features represented by red circles? _____

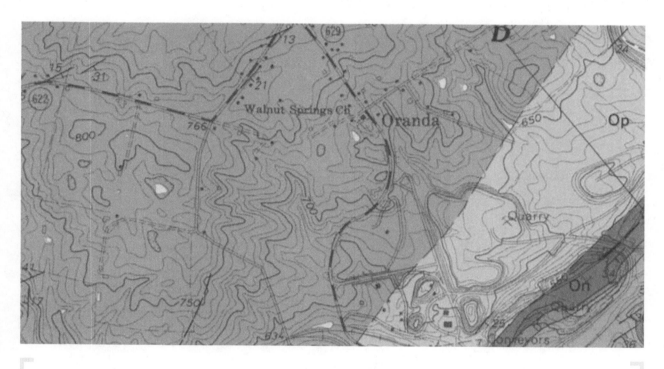

(Source: USGS)

Site 4: Hites Chapel

At Hites Chapel, what is the geological formation symbol? _____ What type of rocks are here? _____

Copy one of the strike and dip map symbols here as it is shown exactly on the map. _____

What is the approximate elevation of the nearby river valley? _____

The nearby hilltop (NE corner of the figure) has a surveyed elevation of 905 ft. What is the local relief? _____.

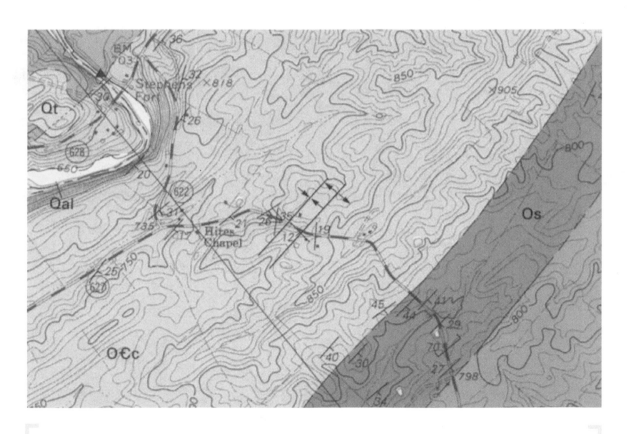

(Source: USGS)

PHYSICAL GEOLOGY LAB EXPLORATION

Site 5: Route 628 - 631

As you continue eastward along route 628-631 crossing Poplar Ridge, what structures do you drive across?

_____

What type of fault do you drive across? _____

Using the online map resource, use the geological cross section from A-A to verify your answer above. Was the fault what you expected based on the symbols?

Referring to the online map resource, in what general direction must you head to return to the town of Reliance? _____.

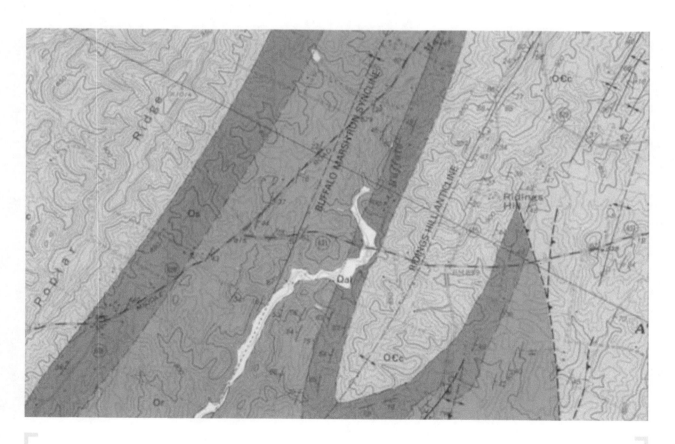

(Source: USGS)

LAB NINE

GLACIATION

LEARNING OBJECTIVES:

- Understand solar energy distribution and albedo/reflectance.
- Learn the two major types of glaciation.
- Understand glacier systems: how glaciers form, move, and mass balance.
- Understand the processes of glacial erosion, transportation, and deposition.
- Be able to identify the common glacial landform features.
- Obtain a basic understanding of climate change and ice age.

Glacier Systems

A glacier system is known as the "river of ice," a moving system of solid ice (see Figure 9.1). Glaciers represent moving ice formed from compacted snow that originates in cold regions.

All glaciers, regardless of their size, share some common properties. Glaciers move under the influence of gravity, and each glacier has two fundamental parts: zone of accumulation and zone of ablation. The snowline marks the dynamic boundary or equilibrium between the two zones.

Zone of accumulation: part of the glacier where there is a net gain of ice from the addition of compacted snow

Zone of ablation: part of the glacier where there is a net loss of ice due to melting, calving, and evaporation

Snowline: the lowest elevation above which snow can stay on the ground throughout the year

Mass balance: the balance between the net gain and net loss of ice

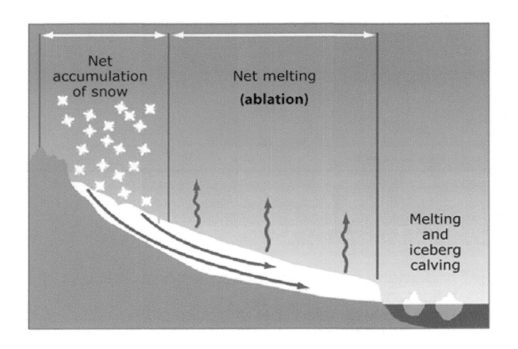

FIGURE 9.1 Schematic profile of a glacier system
(Source: USGS, http://pubs.usgs.gov/fs/2009/3046/)

PHYSICAL GEOLOGY LAB EXPLORATION

1. What is the boundary between the zone of accumulation and the zone of ablation?

2. Does the snowline stay constant over time?

3. Do all places on the earth's surface have the same snowline?

Glacier Development and Solar Energy Distribution

In order for fallen snow to stay year-round, the temperature must be below the freezing point. As such, glaciers only develop in certain locations. Glacier formation and development is largely related to solar radiation.

The solar energy distribution depends on two major factors: the intensity and duration of sunlight received at each location. There are many factors that can affect the energy distribution, such as solar output, atmospheric condition, location, surface land use and land cover, and subsurface activities. To make things simple, we will focus on two of the major factors: the latitude of a location and seasons, which are caused by the relative tilt of earth's axis as it revolves around the sun.

In addition to solar radiation, the earth's surface temperature at a given location is also related to land cover, proximity to the coast, and some other factors. The intensity and duration of solar radiation decrease with latitude. Tropical regions receive more direct sunlight, so the average temperature of the tropics is higher. The temperature also decreases with altitude, and high mountainous regions often exhibit vertical seasonality and climatic zones.

The surface materials of varied land uses and land covers interact with solar radiation differently by reflecting and absorbing different amounts of solar energy. Consequently, the surface temperature at the same location varies with distinct land covers. To better understand the diverse factors and appreciate the formation and movement of glacier systems, let's run a simulation.

Simulation 1: Solar Energy Distribution on the Earth Surface

As discussed in the previous section, the solar energy distribution varies across the earth surface, particularly from the tropics to the poles. To illustrate this point, we will do an experiment to simulate earth–sun dynamics. The purpose is to understand how solar energy is distributed across the earth's surface by using a lamp to simulate the sun and a globe as the earth. To make things easier, we will ignore the 23.5-degree tilt of the earth's axis (see the setup shown in Figure 9.2).

Please measure the intensity of the simulated solar radiation at every 30 degrees using a lux meter and record the readings in the table below.

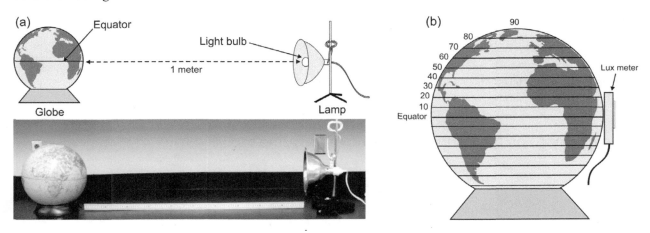

FIGURE 9.2 The intensity variation of simulated solar radiation along the latitude (Diagram credit: Brent Zaprowski)

Latitude (N)	Intensity	% Radiation from equator
0°		
30°		
60°		
90°		

Apply What You Learned

1. Where does the earth surface receive the most solar radiation?

2. Based on this information, where do you expect glaciers to develop? Why?

3. Does the solar radiation decline at a constant rate?

Simulation 2: Albebo of Different Earth Materials

After figuring out why glaciers form, you may wonder how they maintain an energy balance and remain in a solid state for particularly long periods of time. The reason for this is quite elegant and occurs because different earth materials interact with solar energy very differently, in addition to the distribution variations of solar radiation. If a material reflects more sunlight, then it absorbs less and stays cool. Some common types of earth materials include water, vegetation, snow/ice, soil, and rocks. In this experiment, we'll place these materials in different cups and place them on the board shown in Figure 9.3.

Please follow the setup in Figure 9.3, take the measurements, and answer the questions.

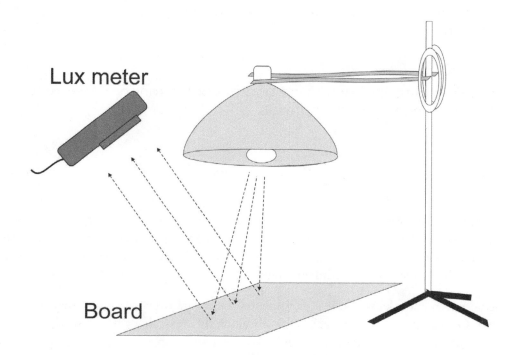

FIGURE 9.3 Simulated reflectance of common earth materials
(Diagram credit: Brent Zaprowski)

Earth Materials	Albedo/Reflectance	Rank (Highest to Lowest)
Water		
Vegetation		
Soil		
Snow/Ice		

1. What earth material has the highest albedo or reflects the most sunlight?

2. Do you think the depth of continental ice sheets is small or great based on this simulation? Explain your answer.

3. Can all environmental conditions remain constant? Will glaciers become larger, smaller, or remain similar in size? Why?

4. Explain possible factors that determine and affect the ranking/order of the relative albedo of the four types of earth materials.

Glacier Movement

Glaciers move! There are two types of glacier movements. The first type is referred to as apparent movement, dictated by climate and, to a lesser degree, by seasonal temperature changes. If the climate gets colder, more snow accumulates and glaciers advance (get larger); if the climate gets warmer, more melting and calving and evaporation occurs; glaciers shrink and retreat.

The second type of movement is known as real movement. Glaciers are rivers of ice, and they move under the influence of gravity. Overall, glaciers move relatively slowly along their base, known as basal slip. Generally, the movement rate is about a foot or so per day. However, a glacier can surge up to tens of feet per day. In addition, not all parts of a glacier move at an equal rate. There are differential movements—glacier ice moves slower along sides and near the bottom due to frictional drag. The differential movement often results in crevasses in glacial ice.

Types of Glaciers

Based on our knowledge and our simulation, we know glaciers develop in cold places. Glaciers are classified into two major types: valley glaciers and continental glaciers.

Valley Glacier

Valley glaciers are also known as alpine glaciers or mountain glaciers. This type of glacier is a system of moving ice that develops in the high mountains and flows downslope along preexisting stream valleys.

The growth of valley glaciers is related to snow accumulation; cirque glaciers grow out of snowfields and gradually extend to valleys and down to the piedmonts, the areas near the foot of a mountain. Therefore, valley glaciers converge and grow into different sizes, like a river system.

Continental Glacier

A continental glacier is also known as an ice sheet since it covers a large area like a continent (see Figure 9.4). It is formed in the polar (high latitude) region. The thickness of a continental glacier can reach 2 to 3 miles. As a result, its flow is not confined to any preexisting river channels.

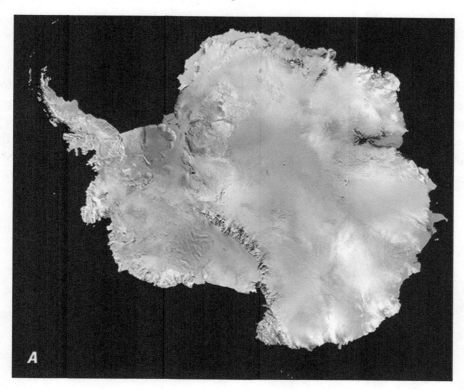

FIGURE 9.4 NOAA AVHRR image mosaic of the Antarctic ice sheet
(Source: http://pubs.usgs.gov/pp/p1386a/gallery2-fig06a.html)

Alpine Glaciation

A glacier is by far the most powerful erosion and transportation agent as it scours and moves along its path.

Glacial erosion: valley glaciers erode the landscape through plucking and abrasion. As snow accumulates, glacier ice develops beneath the snow and often freezes on the bedrock. When it moves, it drags and pulls pieces of rocks along like a bulldozer, which is called **plucking**. The plucked rock materials in turn became abrasive materials and rasp against rocks along the glacier's moving path.

Over time, amphitheater-like depressions called **cirques** form at the head regions of mountain glaciers. Mountain slopes get scoured down by cirques and ice wedging, leaving behind sharp peaks as **horns** and eroded depressions filled with water called **tarns**. The original drainage divide between two valleys is often carved by adjacent valley glaciers, forming a sharp **arête**. In addition, glaciation creates largely straight **U-shaped troughs** or valleys. After the climate gets warmer and glaciers retreat, branch valleys often hang high above the main glacial trough, forming **hanging valleys**. The abrasion create unique polished grooves and shiny rock surfaces. Post glaciation topography is far more sharp edged, pointy, and jagged, strikingly different from the landforms before the glacial period.

Glacial transportation and deposition: Glaciers have enormous power to carry rock debris along their bottom, within ice, and on the top of the glacier. The sediments transported by glacier ice are typically unsorted and known as **till**. The particle size ranges from **rock flour** (fine dust particles) to giant boulders, often known as glacier erratics (see Figures 9.5 and 9.6) which are transported and deposited at a location far away from their origins by glacier.

The sediments deposited by the ice and melting water of valley glaciers form different landform features. **Lateral moraines** form along the sides of a valley glacier. When two valley glaciers converge, two lateral moraines merge and form a **medial moraine**; mounds of rock debris appear in the midst of a glacier. As a glacier reaches its terminus, the moving ice acts like a conveyor belt and deposits its load at the end zone, forming a terminal or **end moraine**. If the climate changes over a period of time, a series of **recessional moraines** will have formed behind the terminal moraine.

FIGURE 9.5 Valley glaciers (Credit: Bruce F. Molnia, USGS, http://water.usgs.gov/edu/gallery/watercyclekids/glacier.html)

FIGURE 9.6 Bridalveil Fall at Yosemite (Credit: Alex Demas, USGS)

Apply What You Learned

1. Please label the glacial landform features in Figures 9.5 and 9.6: horn, arête, lateral moraine, medial moraine, end moraine, hanging valley, U-shaped trough, and cirque.

Continental Glaciation

Unlike alpine glaciation, continental glaciation overwhelms landscapes due to the enormous thickness and erosive power, producing a more leveled landscape, polished and grooved bedrocks, numerous lakes, erratic boulder fields, and large glacial moraines (Figure 9.7).

FIGURE 9.7 Glacial boulders and kettle lakes (Credit: Mara Chen, Yellowstone and Nova Scotia)

Impact of Climate Warming on Glaciation

The previous two simulations showed the uneven distribution of solar radiation and the varied reflection of earth materials to solar energy on earth. Now, consider how global climate change may impact glaciers. Figure 9.8 shows the retreat of Columbia Glacier over the last two decades.

FIGURE 9.8 History of Columbia Glacier retreat (Credit: R.M. Krimmel ,USGS)

1. When a glacier retreats, does the ice move backward?

2. Do you think the retreat of Columbia Glacier represents a local issue or a global one?

3. What additional information would you search for to help prove your opinion?

Simulation 3: Ice Melting under a Lamp

We poured 400 milliliters of water into beakers and put them in a freezer. This simulation is to test how the ice melts under the light (i.e., simulated heat source) of a lamp. At 15-minute intervals, observe the process, dump the melting water into a graduated cylinder, and record the volume of melt water after each 15 minute time interval, and calculate the ice melting rate. After completing the experiment, plot the data in Figure 9.9.

Time (minutes)	Melting Water	Water Left in Ice	Melting Rate %
0	0 ml	400 ml	0
15			
30			
45			
60			

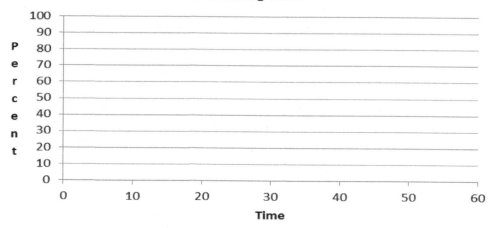

FIGURE 9.9 Ice melting simulation

Apply What You Learned

1. Does the ice melting follow a linear advancing or retreating pattern?

2. If not linear, what is the implication of the impact of climate change on glaciers?

3. Based on the history of Columbia Glacier and the simulation experiment, what do you think about the global sea level change?